M000187568

CLIMATE:
THE COUNTER CONSENSUS

CLIMATE:
THE COUNTER CONSENSUS

A PALAEOCLIMATOLOGIST SPEAKS

Robert M. Carter

STACEY
INTERNATIONAL

Climate: The Counter Consensus

STACEY INTERNATIONAL

128 Kensington Church Street

London W8 4BH

Tel: +44 (0)20 7221 7166; Fax: +44 (0)20 7792 9288

Email: info@stacey-international.co.uk

www.stacey-international.co.uk

ISBN: 978 1 906768 29 4

3 5 7 9 0 8 6 4 2

Printed in Turkey by Mega Basim

CIP Data: A catalogue record for this book is available from the British Library

*To the scientists, technicians, administrative staff
and ships' and drilling crews of the
Deep Sea and Ocean Drilling Programmes,
whose exertions have helped to unlock
Planet Earth's archive of climate change.*

The idea that human beings have changed and are changing the basic climate system of the Earth through their industrial activities and burning of fossil fuels – the essence of the Greens' theory of global warming – has about as much basis in science as Marxism and Freudianism. Global warming, like Marxism, is a political theory of actions, demanding compliance with its rules.

Marxism, Freudianism, global warming. These are proof – of which history offers so many examples – that people can be suckers on a grand scale. To their fanatical followers they are a substitute for religion. Global warming, in particular, is a creed, a faith, a dogma that has little to do with science. If people are in need of religion, why don't they just turn to the genuine article?

Paul Johnson

CONTENTS

LIST OF FIGURES

Prefatory Essay

The author of *Climate: the Counter Consensus* and I disputed over the book's title. Professor Robert M. (Bob) Carter resisted the notion that anything he wrote might be taken to imply that science could ever be about 'consensus': science was about verifiable fact. I saw his point. But I argued that the popular 'consensus' was indeed that science was 'about' *consensus*; here was the democratic virus going about its business. Hence, precisely, said I, was how so much of the world, spoken for by the (largely democratically attuned) G8, had allowed itself to be aroused to a frenzy of alarm at climate change. The 'consensus' among scientists was that global warming was taking place, that it was an imminent threat to the 'survival of the planet', and that it was attributable wholly or at least significantly to man-generated emissions of carbon dioxide.

I was aware that that very statement of mine contained four *factual* mis-assumptions, including that scientific truth could ever be determined by 'consensus'. As others have surely pointed out, the prevailing 'consensus' among astronomers in the early seventeenth century was that the sun circled the Earth: Galileo was locked up for upholding the reverse.

The title of your book, I could assure our author, countered not merely the supposed consensual opinion of collective scientific peers but the validity of the very concept of a 'consensus' of scientific opinion.

As we are all now well aware, and Bob Carter makes clear in this work, such 'consensus' as may have been thought to prevail among the scientific community concerning the warming of the globe was sustained throughout by data assembled by a small clique of well-placed and often lately-arrived climatologists who have been selective, slovenly or wilfully distortive in their evaluation of it. *They* have led the dance on this issue – a dance joined, I need not stress, since the 2008 Presidential election in the US, by the leaders of all but one of the democratic West, and all but one of the official Oppositions, the exceptions being, respectively, the economist Vaclav Klaus, President of the Czech Republic, and (somewhat shyly) Australia's Liberal-National Party opposition. There are symptoms at the time of my writing this of America's Republicans adopting a similar stance.

All others were responding, oblivious of cost and consequence, to what they and, most persuasively, their electorates had been gulled into accepting as scientific truth concerning anthropogenic global warming. The clique itself clung to its tenets for dear life, for it prospered from the ardent funding of their institutions, programmes, and university departments, and the fame and influence of the protagonists themselves.

An absence of academic discipline and personal scruple in this field of fast expanding international significance was evident early, not least to the present author. In the late 1980s the economist Professor Lester Lave of the Carnegie Mellon University of Pittsburgh, giving evidence to the Senate Committee on Energy and Natural Resources on the 'controversiality' of the theory of anthropogenic emissions of CO_2 affecting global temperature, was summarily silenced by Senator Al Gore. In 1992, Dr Richard Lindzen, America's pre-eminent atmospheric physicist, wrote a paper warning of the extraordinary pressure to stifle dissent or even debate on the issue. Meanwhile, alarmist predictions multiplied, were widely upheld, and left unchallenged. They were emanating, after all, from bodies or organizations of apparent authority and presumed objectivity, including of course the Intergovernmental Panel on Climate Change (IPCC), established in 1988 by two organs of the United Nations, at the instigation of the Swedish meteorologist Bert Bolin. Alongside the IPCC, as a prime source of its opinions, were the Hadley Centre for Climate Prediction, simultaneously established at the instigation of Margaret Thatcher as an adjunct to the UK Met Office, and the Climatic Research Unit of the University of East Anglia. The IPCC's 31-person directorate, currently headed by the Indian businessman and economist Dr Rajendra Pachauri, was drawn not from scientists as such but from a miscellany of senior civil servants, academics and savants from a politically correct span of nationalities and continents.

Human responsibility for global warming rapidly took centre stage in the workings of the IPCC, its pronouncements, protocols and successive 'Earth Summits' at Rio de Janeiro, Kyoto, Bali, Bonn and Copenhagen between 1992 and 2009. Worldwide sentiment was mobilized by the traditional 'green' movements such as Friends of the Earth, Greenpeace and the World Wildlife Fund. There was a promise here of votes. The world's most gleaming politicians joined Al Gore in his declaration that 'the time for debate is over'. 'This disaster,' declared Tony Blair in 2006, 'is not set to happen in some science fiction future, many years ahead, but in our own lifetime.' 'The science is beyond dispute,' confirmed Barack Obama, campaigning for the Presidency against a Republican Party still tainted by scepticism, 'and the facts are clear.' So clear were the facts to the Chief Negotiator of the G77 (group of developing nations, including China) at the Copenhagen Earth Summit of 2009, Lumumba Stanislaus Di-Aping, from Sudan, that he declared the $100 billion being offered to fund the containment of his members' carbon emissions and adapting to change 'would not be enough to buy the poor nations the coffins' for those swamped in their island states and facing 'certain death' in an Africa condemned to 'absolute devastation'.

Let us recall that the science of anthropogenic global warming (AGW) had been fertilized at its inception by the Green-ish, New Age-ish miasma of the Sixties, and its aura of ideological anarchism. The 'consensus' was visceral and visionary, even apocalyptic: to question or challenge its tenets was an emotional and, quite soon, ideological affront. Young thinkers at climate and economic summits expressed their protests by smashing windows of global structures, be they banks or restaurants, winning honour in the name of 'saving the planet' from catastrophic warming. The somewhat hermetic doctrines of the quasi-Marxist Jerome Ravetz, proposing the right of what he called 'post-normal science' to manipulate scientific findings for social purposes, attained intellectual fashionability.

Yet from the early 1990s, especially in North America, a few voices of informed dissent have persistently and intelligently challenged the scientific basis of the alarmist consensus. By around 2005, the number of well qualified 'climate sceptics' had swelled considerably and their voices were beginning to be heard and even heeded. Among them was that of Professor Bob Carter, doyen of that rare international species, the palaeoclimatologist, whose discipline is central to the truth on this issue. Given the measure of stifling, and indeed intimidation, of the scientific community, dissent was perhaps heard more from those of other disciplines, especially economics and statistics. Thatcher's esteemed Chancellor of the Exchequer and former Secretary of State for Energy, Nigel Lawson, was an early voice counselling caution: he remains a force for clarity of thought and analysis. Elsewhere in Europe was his fellow statesman Vaclav Klaus, who like Lawson wrote his own book on the subject, and the distinguished scientist and France's former Minister for Education, Claude Allègre. The list of expert dissent in the first decade of the century prominently includes the names of the Canadian statistician and mining financial analyst, Steve McIntyre; the French engineer Christian Gerondeau, whose work $CO_2 - Un Mythe Planétaire$ we are about to publish in an updated edition as *Climate: the Great Delusion*; and the economists Ross McKitrick and David Henderson, formerly senior statistician at the Organisation for Economic Co-operation and Development (OECD). Carter's fellow Australian, Ian Plimer, Professor of Mining Geology at the University of Adelaide, has valuably authored *Heaven and Earth*.

Reaching the sophisticated public through the media have been the campaigning author and journalist Christopher Booker;

the researcher Andrew Montford, author of *The Hockey Stick Illusion*, recently published to wide acclaim in this same Stacey International series, Independent Minds, consolidating his formidable following in the blogosphere. Several influential journalists in, at least, the British and American press were since 2009 beginning to perceive a monstrous deception at work. Among them (in Britain) is the feisty polemicist James Delingpole who dubbed as Climategate the exposure of the doctoring of the data by East Anglia University's Climatic Research Unit by emails leaked in December 2009. In Australia, the opinion writer Andrew Bolt has created a much visited blog which provides almost daily excoriation of global warming propaganda.

Professor Carter lays out in the present succinct yet comprehensive work a scenario which future generations will regard as a period of collective insanity. Investigations at that future time will be concerned not with climate change but with the Dionysiac delusion of a style and magnitude comparable to that which induced (for instance) the mass fervour for the promise of fascism in Italy and Germany in the 1920s and '30s, such as included within its range of dupes or fellow-travellers many of the *cognoscenti* of the period. We are already three generations beyond that period yet still await our serious historians to delve the full answer as to *how it could have been so*.

Perennially, mankind has been drawn to visions of eschatological extinction. Symptoms of this same instinct at work are to be seen in the allure of Armageddon, in the carefree abandonment to death-or-glory on the entry of nations into war and the concomitant plummeting of suicide rates. Catastrophilia is ever with us, accompanied by wild-eyed summons to action. Let me cite personal experience. In 1972 my eponymous publishing house was bringing out an up-to-the-minute series of books under the collective title 'Prospect for Man', mostly by respected environmentalists of the day, in the face of alarm at a comparable imminent catastrophe.

Our flagship title was the 170-page *Blueprint for Survival*. Over two million copies were sold or distributed. It was written by the editor of the *Ecologist*, Edward Goldsmith, and four others, and listed in the opening pages were 38 of Britain's most honoured scientists, economists and environmentalists who endorsed the work, including 18 Professors, two Nobel laureates, and seven Fellows of the Royal Society. The jacket explained that it concerned 'our imminent future, which individuals and governments can ignore only at their peril'. Armageddon was forecast well ahead of the year 2000, by which date, incidentally, hydrocarbon fuel sources would be exhausted as well as the world's copper, mercury, molybdenum, nickel, lead, platinum, zinc, silver and gold. There would be extensive desertification around the world, since the supply of cultivable land would have been exceeded by demand for it. As publisher, I wrote the Foreword, in which I opined 'the publication of a *Blueprint for Survival* will prove in years to come to have marked a turning point in attitudes which will affect the course of our civilisation.' The cause, however, of this impending catastrophe was not global warming: it was over-population.

We were fooled, were we not? We had got the science and demographics ridiculously wrong, the Nobel laureates, Fellows of the Royal Society, and the rest of us.

Edward Goldsmith's *Ecologist* magazine, supported by his brother Jimmy (the late Sir James), my old school friend, has come in due course to be edited by Jimmy's son Zac who for better or for worse is environmental adviser to David Cameron, Britain's Conservative leader. May my younger friend David take caution. These Goldsmiths are highly plausible.

The difference between that earlier and half-forgotten surge of alarmism and today's is in the more emphatically *religious* character of our present movement. There is the ethical dimension. As with over-population global warming is an issue to be given not only close academic attention but political action too; but the postulation of willed anthropogenic emissions of carbon dioxide being responsible for irreversible warming carries an implication of blame and attendant guilt. Given the ideological tilt of the Green movement, the massive pollution of monolithic industry in socialized countries has been overlooked. The culprit immediately to hand was capitalism and 'big business'. *Green*, in Lawson's aphorism, became the new *red*. New Age sentiment sanctifying a vague return to nature and Gandhian craft, espoused the alarmism, especially in the burning of fossil fuels. John Houghton, a former Chair of the IPCC and a Fellow of the Royal Society, had purportedly been overheard passing the word around, 'Unless we announce disasters, no one will listen', and were it he to have uttered those words (for he has energetically denied it) they were surely listened to assiduously. The tendentious film to which Al Gore had given his name set the pulses racing. Even the Churches joined in. Man's greed, ran the rune, was about to destroy God's Earth.

A factor of what I venture to call genuine religion had come into play, in that it is present in the Abramic traditions. We recall Adam who in eating of the Tree of the Knowledge of Good and Evil attained to consciousness and chose to disobey; and hence his inherent sense of original sinfulness. For although Man may feel himself to be made in the image of God as the Book of Genesis avers, he at once discovers he cannot *be* God.

Instead, utopian fantasies possess and beguile the ever-seducible human psyche: in the century just past, most obviously and calamitously, Marxism – an *aperçu*, I may mention, of Dr Benny Peiser, the social anthropologist who is at present a colleague of Nigel Lawson on the latter's Global Warming Policy Foundation. That utopian brand outflanked and outlasted fascism, albeit not by all that much in the longer view of history. Into the ensuing world of more-or-less godless consumerism, the sense of sinful inadequacy has ineluctably persisted. The deeper live-ability of life, one daresay *meaning*, has remained teasingly elusive. The young of the developed world, so-called, and especially the idealistic, find themselves with the need to ascribe a sense of internal smear to something or someone. Whom shall they blame, our innocents? Who must now wipe the planet clean of Man's carbon footprints? Why, the racketeers who have constructed the blighted world the innocent have been saddled with: the selfish and the greedy. By demonising others, the blight of guilt is eased.

To be fair to the Christian churches, in particular the Anglican, the dualism between creation and creation's dominant species, Man, is a heresy justly perceived: it has been on the prowl since Eden. But Canterbury's Archbishop, whom I admire, and who spoke at the Copenhagen summit, is found to be astride the wrong horse in the present somewhat fantastical guise which the contra-dualist contest appears. Christians are more readily gulled than most, and forgivably so.

This factor of encircling guilt, in which naturally the Green protesters are themselves complicit, is underlain by a deeper neurosis of our aeon, namely the presumed chasm of differentiation of Man from the Cosmos, such as has laid upon the human race a Manichaeistic obligation to exert its will upon the totality of the creation in which he exists. A characteristic of the Earth Summit at Copenhagen in December 2009 was the dismaying *hubris* by which such a politician as Gordon Brown could presume to promise that he would see to it that the rise in global temperatures would be restricted to 1.5 degrees Celsius. Not even the courtiers of King Canute would have suppressed a smirk. Yet the presumption still prevails that if the climate of the world is going awry, it must be Man's doing, and that it is for Man to rescue himself from his own folly.

That there is no evidence of the climate going awry in the longer view of climate history is what Professor Carter sets forth in this work. This is a proposition offensive to many since it removes from them what has become an alleviation of the neurosis – that is, removes the justification to relieve their opaque sense of guilt by loading guilt on others. The invented chimera of anthropogenic global warming, clutched at by the psyche, is in danger of being snatched away.

Indeed so, by such as Bob Carter. The self-declared innocents are now to learn they have been betrayed by their prophets, who have dissembled, told half-truths, cherry-picked their data, fantastically exaggerated, and suppressed the circulation of better science. A great cloud of doubt and disillusion lowers over the entire issue on which the fate of the planet was supposed to hang.

At last the scientists with the right to be heard are writing for the general reader and for the common voters. Outstanding among them is Professor Carter, author of the present work. No other palaeoclimatologist stands above him in the range, precision of knowledge, and ability to communicate it. He writes with balance, humour and caution, and the courage to define the boundaries of both the known and unknown. But he knows the sophisticated world has been massively deceived. This work tells the measured truth of that deception.

How shall it all turn out? The vastly ramified financial edifices of carbon trading, inflated subsidies for essentially wasted sources of 'renewable' energy, the brokers and middlemen, the bankers' ramps, the existing and impending carbon taxes levied not only nationally but by multi-lateral treaty, the subsidized scourge of biofuels production so devastating to creation's diversity in the rainforests – what shall become of it all? The voters are getting to know: a potential democratic self-heal. We publish this work at that point where that decisive player in the drama, the electorate, is ready to wake up and face up to the truth. They will awake to impositions of formidable public expenditure which they know to be futile. Something has to give.

Tom Stacey
March 2010

Acronyms and abbreviations

AEF – Australian Environment Foundation
AMO – Atlantic multi-decadal oscillation
AP Index – average planetary magnetic index
CERN – European Organisation for Nuclear Research
CET – central England temperature index
CRU – Climatic Research Unit (University of East Anglia)
CSIRO – Commonwealth Science and Industrial Research Organization
    (Australia)
EMA – Emergency Management Australia
ENSO – El Nino-Southern Oscillation
EPA – Environment Protection Agency (US)
FEMA – Federal Emergency Management Agency (US)
GBR – Great Barrier Reef
GCM – general circulation model
Gt – Gigatonne
Gt C/yr – Gigatonne of carbon per year
IMO – International Meteorological Organization
INCCCA – Indian Network on Comprehensive Climate Change Assessment
IPCC – Intergovernmental Panel on Climate Change (United Nations)
    1AR – First Assessment Report by IPCC (1990)
    2AR – Second Assessment Report by IPCC (1996)
    3AR – Third Assessment Report by IPCC (2001)
    4AR – Fourth Assessment Report by IPCC (2007)
ky, ka – thousand years, thousand years ago
LIA – Little Ice Age
LOD – length of day
LRSL – local relative sea level
MSL – mean sea level
MSU – microwave sensing unit
mW/m^2 – milliwatts per square metre
My, Ma – million years, million years ago
NERC – National Environmental Research Counci (UK)l
NOAA – National Oceanic and Atmosphere Administration (US)
PDO – Pacific Decadal Oscillation
ppm – parts per million
RSS – Remote Sensing Systems
TSI – total solar irradiance
UAH – University of Alabama, Huntsville
UNEP – United Nations Environment Programme
UNFCCC – United Nations Framework Convention on Climate Change
W/m^2 – watts per square metre
WMO – World Meteorological Organization

Author's Preface

Climate change knows three realities. *Science reality*, which is what working scientists deal with on a daily basis. *Virtual reality*, which is the wholly imaginary world inside computer climate models. And *public reality*, which is the socio-political system within which politicians, business people and the general citizenry work.

The *science reality* is that climate is a complex, dynamic, natural system that no one wholly comprehends, though many scientists understand different small parts. So far, and despite the very strong public concern, science provides no unambiguous evidence that dangerous or even measurable human-caused global warming is occurring. Second, the *virtual reality* is that computer models predict future climate according to the assumptions that are programmed into them. There is no established Theory of Climate, and therefore the potential output of all realistic computer general circulation models (GCMs) encompasses a range of both future warmings and coolings, the outcome depending upon the way in which a particular model run is constructed. Different results can be produced at will simply by adjusting such poorly known parameters as the effects of cloud cover. Third, *public reality* is that, driven by strong environmental lobby groups and evangelistic scientists and journalists, to whom politicians in turn respond, there was a widespread but erroneous belief in our society in 2009 that dangerous global warming is occurring and that it has human causation.

The regular occurrence around the world of natural climate or climate-related disasters such as storms, floods, droughts and bushfires makes it self-evident that all countries, be they Western or third-world nations, need to possess sensible policies to deal with national climate hazard. Furthermore, such policies need to be tailored to the particular risk environment in each country or large region (for instance, alert to typhoons in Japan and bush wildfires in California) rather than tailored to some amorphous 'global climate'; no-one, but no-one, lives in a global climate. Yet expensive television advertisements run in 2009 by, for example the British and Australian governments, make it clear that their current 'climate policy' is concerned with addressing the virtual reality of hypothetical human-caused global warming rather than the actual reality of everyday climate variability. In truth, Western nations don't have national climate policies at all, but rather imaginary global warming policies instead.

The current public 'debate' on climate is not so much a debate as it is an incessant and shrill campaign to scare the global citizenry into accepting dramatic changes in their way of life in pursuit of the false god of preventing dangerous global warming. Furthermore, this debate is persistently misrepresented by the media as being between morally admirable 'believers' and morally challenged 'deniers'. In reality, such shallow moralities have nothing to do with science, which derives its own considerable moral and practical authority from the objective use of facts, experiments and analytical reasoning to test hypotheses about the natural world.

It is widely believed, and wrongly, that the study of climate change is the exclusive province of meteorologists and climatologists. In reality, scientists who study climate change come from a very wide range of disciplines that can be grouped into three main categories. The first group comprises scientists who are expert in meteorology, atmospheric physics, atmospheric chemistry and computer modelling, who mostly study change over short periods of time, and are primarily concerned with *weather processes* (and, by extension, climate processes); a second group comprises geologists and other earth scientists, who hold the key to delineating *climate history* and the inference of ancient climate processes; finally, a third category comprises those persons who study enabling disciplines like mathematics, statistics and (perhaps) engineering.

In this context, competent scientists from all these three groups accept, first, that global climate has always changed, and always will; second, that human activities (not just carbon dioxide emissions) definitely affect local climate, and have the potential, summed, to measurably affect global climate; and third, that carbon dioxide is a mild greenhouse gas. The true scientific debate, then, is about none of these issues, but rather about the sign and magnitude of any global human effect, and its likely significance when considered in the context of natural climate change and variability.

As a generalization, it can be said that most of the scientific alarm about dangerous climate change is generated by scientists in the meteorological and computer modelling group, whereas many (though not all) geological scientists see no cause for alarm when modern climate change is compared with the climate history that they see every time they stand at an outcrop, or examine a drill core. Of course, attaining a full perspective on climate change requires at least a passing familiarity with all of the three groups of disciplines, a demand that tests even the most polymathic of the scientific brethren. The fact that scientific opinion is divided over the global warming issue is therefore not unusual, and in part follows inevitably from the diversity of knowledge involved; discussion and rational argument are the lifeblood of science, and are indicative of a healthy rather than unhealthy state of affairs. Unlike policy, science is never 'settled'.

In this book I will describe the natural variations in climate that we are heir to, examine the possibility of an additional and measurable human effect, explain why carbon dioxide taxation is a non-solution to a non-problem, and finally show how a cost-effective and prudent climate policy can be included within national plans that address all major climatic hazards.

Chapters 1-6 outline the science of the climate change issue, including a discussion of the vexed virtual realities of GCM computer modelling. Chapters 7-10 contain a discussion of the powerful social and political forces that are still calling for action against global warming at a time when the globe has actually been cooling for a decade. Chapter 11 identifies the forward path, which should be preparation for, and adaptation to, climate change as it happens irrespective of its causation. For many of the greatest human disasters are caused by natural climatic events, and it is self-evident that we need to handle them better. At the same time, it is simply hubris to imagine that our present understanding of planet Earth is adequate to allow us to successfully engineer future climate. Finally, Chapter 12 describes briefly the breaking of the Climategate scandal in November 2009 and the closely following IPCC climate summit meeting in Copenhagen in December, and Chapter 13 (*Postscriptum*) presents a final and balanced summary statement about the possible human influence on global climate.

My aim in this book has been not only to create an alternative narrative for the 1988-2009 global warming story. Rather, I wish also to encourage people to trust authority less and their own brains more as they assess the likely dangers of both known natural and hypothetical human-caused global climate change. Towards that end, as well as to enhance readability of the main text, most of the technical detail is provided in the sources listed in the end-notes, which provide many independent references to published papers, articles and high-quality web commentaries. Consulting these sources is rewarding in its own right, and it is also an excellent antidote for those who hitherto have heard only the 'authoritative' views of vested interest organizations such as the United Nations, government science agencies and national science academies.

Climate is, and will continue to be, created and controlled by immense and complex natural forces, not by political fiat. Any practical way forward out of the present 'stop global warming' fiasco must acknowledge that reality, as does the adaptive policy, Plan B, outlined in Chapter 11 of this book.

Introduction

Reality is only an illusion, albeit a very persistent one.

(Albert Einstein)

To effectively communicate, we must realize that we are all different in the way we perceive the world and use this understanding as a guide to our communication with others.

(Tony Robbins)

Before human-caused global warming[1] can become an economic, social or environmental problem, it first has to be identified by scientific study as a dangerous hazard for the planet, distinct from natural climate change.

This notwithstanding, several distinguished economists have recently written compendious papers or reports on the issue, for example the UK's Nicholas Stern[2], USA's William Nordhaus[3] and Australia's Ross Garnaut[4]. These persons, and many other public commentators and politicians as well, have naively accepted that there is a scientific consensus (the phrase itself being an oxymoron) that dangerous, human-caused global warming is occurring, as set by the views and advice of the Intergovernmental Panel on Climate Change (IPCC)[5].

The IPCC is the United Nations body that in 1995 allowed a single activist scientist, Ben Santer, to rewrite parts of the key Chapter 8 (*Detection of Climate Change and Attribution of Causes*) of its Second Assessment Report in alarmist terms, changing a previous wording that had been agreed among the other scientific authors. The rewriting was undertaken in order to make the chapter agree with a politically contrived statement in the influential Summary for Policymakers, to whit 'the balance of evidence suggests a discernible human influence on global climate'. This statement being the opposite of the conclusion drawn in the original Chapter 8 text, it was obvious from that point onward that IPCC pronouncements needed to be subjected to independent critical analysis. Instead, the opposite has happened and increasingly the world's press and politicians have come to treat IPCC utterances as if they were scribed in stone by Moses. This is a reflection, first, of superb marketing by the IPCC and its supporting cast of influential environmental and scientific organizations (not to mention the bucket-loads of money that have been available in their support[6]); second, of strong media bias towards alarmist news stories in general, and global warming political correctness in particular; and, third, of a lack of legislators and senior bureaucrats possessed of a sound knowledge of even elementary science, coupled with a similar lack of science appreciation throughout the wider electorate – our societies thereby having become vulnerable to frisbee science, or spin.

Having decided around the turn of the twentieth century that '*the science was settled*', for the IPCC said so, politicians in industrialized societies and their economic advisers started to implement policies that they assured the public would '*stop global warming*', notably measures to inhibit the emission of the mild greenhouse gas carbon dioxide into the atmosphere. However, the acronym GIGO (garbage in, garbage out) that has long been applied to computer modelling endeavours applies also to economic studies that purport to give policy advice against the threat of future climate change. For the reality is that no-one can predict the specific way in which climate will change in the future, beyond the general statement that multi-decadal warming and cooling trends, and abrupt climatic changes, are all certain to continue to occur. It is also the case that the science advice of the IPCC is politically cast, and thereby fundamentally flawed and unsuitable for use in detailed economic forecasting and policy creation. This is why Stern's work, for example, has been able to be so severely criticized on both scientific and economic grounds[7], with respect to which the critical essays of Melbourne climate analyst John McLean[8] provide searing insights into the unreliability of the IPCC.

MIT atmospheric physicist Richard Lindzen famously remarked of global warming alarmism a few years ago that 'The consensus was reached before the research had even begun.' Another distinguished natural scientist, the late Sir Charles Fleming from New Zealand, made a similarly prescient statement when he observed in 1986 that 'Any body of scientists that adopts pressure group tactics is endangering its status as the guardian of principles of scientific philosophy that are worth conserving.'

These quotations are apposite, because pressure-group tactics in pursuit of a falsely claimed consensus are now the characteristic *modus operandi* of the IPCC-led global warming alarmists who surround us at every turn. The recent sensational public exposure of email exchanges between climate scientists at the UK's Climatic Research Unit (an organization closely linked with the Meteorological Office's Hadley Centre) and their colleagues around the world has revealed the malfeasance involved for the whole world to see (Chapter 12).

The realities of climate change

Science reality
My reference files categorize climate change into more than one hundred subdiscipline areas of relevant knowledge. Like most other climate scientists, I possess deep expertise in at most two or three of these subdisciplines. Chris Essex and Ross McKitrick have observed[9]:

> Global warming is a topic that sprawls in a thousand directions. There is no such thing as an 'expert' on global warming, because no one can master all the relevant subjects. On the subject of climate change everyone is an amateur on many if not most of the relevant topics.

It is therefore a brave scientist who essays an expert public opinion on the global warming issue, that bravery being always but one step from foolhardiness. And as for the many public dignitaries and celebrities whose global warming preachings fill out our daily news bulletins, their enthusiasm for a perceived worthy cause greatly exceeds their clarity of thought about climate change science, regarding which they are palpably innocent of knowledge.

In these difficult circumstances of complex science and public ignorance, how is science reality to be judged? This question was first carefully thought through in the late 1980s by the senior bureaucrats and scientists who were involved in the creation of the United Nations's IPCC. Key players at the time were Bert Bolin (Sweden), John Houghton (UK) and Maurice Strong (Canada), the two former persons going on to become Chairman of the IPCC and Chairman of Working Group 1 (science), respectively. The declared intention of the IPCC was to provide disinterested summaries of the state of climate science as judged from the published, refereed scientific literature. Henceforward, in the public and political eye, science reality was to be decided by the authority of the IPCC. Accordingly, in four successive Assessment Reports in 1990, 1996, 2001 and 2007 the IPCC has tried to imprint its belief in dangerous human-caused warming on politicians and the public alike, steamrollering relentlessly over the more balanced, non-alarmist views held by thousands of other qualified scientists. Inevitably, and despite the initial good intentions, what started in 1988 as a noble cause had by the time of the 2007 Fourth Assessment Report degenerated into a politically driven science and media circus.

As Chris Essex and Ross McKitrick have written[10]:

> We do not need to guess what is the world view of the IPCC leaders. They do not attempt to hide it. They are committed, heart and soul, to the Doctrine [of dangerous human-caused global warming]. They believe it and they are advocates on its behalf. They have assembled a body of

evidence that they feel supports it and they travel the world promoting it.

There would be nothing wrong with this if it were only one half of a larger exercise in adjudication. But governments around the world have made the staggering error of treating the IPCC as if it is the only side we should listen to in the adjudication process. What is worse, when on a regular basis other scientists and scholars stand up and publicly disagree with the IPCC, governments panic because they are afraid the issue will get complicated, and undermine the sense of certainty that justifies their policy choices. So they label alternative views 'marginal' and those who hold them 'dissidents'.

The basic flaw that was incorporated into IPCC methodology from the beginning was the assumption that matters of science can be decided on authority or consensus; in fact, and as Galileo early showed, science as a method of investigating the world is the very antithesis of authority. A scientific truth is so not because the IPCC or an Academy of Science blesses it, or because most people believe it, but because it is formulated as a rigorous hypothesis that has survived testing by many different scientists.

The hypothesis of the IPCC was, and remains, that human greenhouse gas emissions (especially of carbon dioxide) are causing dangerous global warming. The IPCC concentrates its analyses of climate change on only the last few hundred years, and has repeatedly failed to give proper weight to the geological context of the short, 150-year long instrumental record. When viewed in geological context, and assessed against factual data, the greenhouse hypothesis fails. There is no evidence that late twentieth century rates of temperature increase were unusually rapid or reached an unnaturally high peak; no human-caused greenhouse signal has been measured or identified despite the expenditure since 1990 of many tens of billions of dollars searching for it[6]; and global temperature, which peaked within the current natural cycle in the warm 1998 El Nino year, has been declining since then despite continuing increases in carbon dioxide emission.

Recognition of the post-1998 cooling has been strongly resisted by warming alarmists since it first became evident around 2006, despite which acknowledgement of the cooling has now spread to mainstream journals such as *Geophysical Research Letters* (GRL). A recent GRL paper by Judith Perlwitz and co-authors[11] refers to 'A precipitous drop in North American temperature in 1998', and continues that 'Doubts on the science of human-induced climate change have been cast by recent cooling. Noteworthy has been a decade-long decline (1998-2007) in globally averaged temperatures from the record heat of 1998.' In support of this statement, Perlwitz cites another GRL paper by Easterling and Wehner[12], who, whilst acknowledging the cooling, put a brave face on the matter by concluding that 'climate over the twenty-first century can and likely will produce periods of a decade or two where the globally averaged surface air temperature shows no trend or even slight cooling in the presence of longer-term warming.' It is clearly difficult for even the most straightforward of facts to shift the fierce belief in human-caused warming that is held by these and many other scientists.

In summary, the science reality in 2009 was that the IPCC's hypothesis of dangerous, human-caused global warming had been repeatedly tested and failed. In contrast, the proper null hypothesis that the global climatic changes that we observe today are natural in origin has yet to be disproven (Chapter 6). The only argument that remains to the IPCC – and it is solely a theoretical argument, not evidence of any kind – is that their unvalidated computer models project that carbon dioxide driven dangerous warming will occur in the future: just you wait and see! It is therefore to these models that we now turn.

Virtual reality

The general circulation computer climate models (GCMs) used by the IPCC are deterministic. Which is to say that they specify the climate system using a series of mathematical equations that are derived from the first principles of physics. For many parts of the climate system, such as the behaviour of turbulent fluids or the processes that occur within clouds, our incomplete knowledge of the physics requires the extensive use of parameterisation (read 'educated guesses') in the models, especially for the many climate processes that occur at a scale below the 100-300 km^2 size of the typical modelling grid.

Not surprisingly, therefore, the GCMs used by the IPCC have not been able to make successful climate predictions, nor to match the expected 'fingerprint' of greenhouse gas-driven temperature change over the late twentieth century. Regarding the first point, none of the models was able to forecast the path of the global average temperature statistic as it elapsed between 1990 and 2006. Regarding the second, GCMs persistently predict that greenhouse warming trends should increase with altitude, especially in the tropics, with most warming at around 10 km height; in contrast, actual observations show the opposite, with either flat or decreasing warming trends with increasing height in the troposphere[13].

The modellers themselves acknowledge that they are unable to predict future climate, preferring the term 'projection' (which the IPCC, in turn, use as the basis for modelled socio-economic 'scenarios') to describe the output of their experiments. Individual models differ widely in their output under an imposed regime of doubled carbon dioxide. In 2001 and 2007, the IPCC cited a range of 1.8-5.6°C and 1.4-5.8°C warming by 2100, respectively, for the model outputs that they favoured, but this range can be varied further to include even negative outputs (i.e. cooling) by adjustment of some of the model parameters. Indeed, the selected GCM outputs that IPCC places before us are but a handful of visions of future climate from among the literally billions of alternative future worlds that could be simulated using the self-same models.

It is clear from all of this, and from the more detailed discussion in Chapter 5, that climate GCMs do not produce predictive outputs that are suitable for direct application in policy making; it is therefore inappropriate to use IPCC model projections for planning, or even precautionary, purposes, as if they were real forecasts of future climate. Notwithstanding, it remains the case, amazingly, that the IPCC's claims of a dangerous human influence on climate now rest almost solely on their unrealistic, invalidated GCM climate projections. Which makes it intriguing that during recent planning for the next (5th) IPCC assessment report, due in 2015, senior UK Hadley Centre scientist, Martin Parry, is reported in a *Nature* article as saying: 'The case for climate change, from a scientific point of view, has been made. We're persuaded of the need for action. So the question is what action, and when.'

Well, the IPCC may be so persuaded, but the key question, of course, is what about the rest of us?

Public reality

The answer to that question is that opinion polls in 2007 and 2008 showed that most of the rest of us had become severely alarmed about the threat of human-caused climate change[14]. Therefore, public reality, as perceived until recently by most Western governments, is that their electorates have been expecting them to 'do something' about global warming, i.e. to introduce a carbon dioxide taxation system. Despite rapid swings in public opinion towards less alarmist beliefs in late 2009, it remains the case that there exists a strong disjunction between climate alarm as perceived by the public (strongly egged on by the press) and the science justification for that alarm. How come?

The means by which the public has been convinced that dangerous global warming is occurring are not subtle. The three main agents are the reports from the IPCC that I have already described; incessant bullying by environmental NGOs (such as

Greenpeace, World Wide Fund for Nature, Australian Conservation Foundation, Pew Foundation) and their allied scientists, and by science organizations, political groups and business; and the obliging promulgation of selectively alarmist climate information by the media. Indeed, the combined alarmist activities of the IPCC, crusading environmental NGOs, some individual leading climate scientists and many science agencies and academies can only be termed a propaganda campaign. However, because all of these many interest groups communicate with the public primarily through the gatekeepers of the press, it is the press that carries the prime responsibility for the unbalanced state of the current public discussion and opinion on global warming.

Note on language

Language is the essence of communication. Much science is couched in precise language that appears as jargon, even to well-educated non-scientists. Beyond that, where science bears on environmental issues, choice of language becomes more complex still because of its deliberate politicisation by special-interest groups. Thus phrases such as nuclear waste dump, alternative energy, dirty power, green power, carbon footprint, and more, have been skilfully coined and deployed precisely in order to influence the terms of the social debate.

Sometimes entirely new and inaccurate terminology is involved, as for example when the emotional phrase 'acidification of the ocean' appeared in the early 1990s. Earlier scientific papers related to this topic had carried prosaic, descriptive titles such as: 'The effect of increases in the atmospheric carbon dioxide content on the carbonate ion concentration of surface water at $25°C$' (paper in *Limnology & Oceanography*, 1975), but soon an altogether more portentous style of title emerged, viz: 'Impacts of ocean acidification on marine fauna and ecosystems processes' (paper in *Journal of Marine Science*, 2008).

Scarcely surprisingly, the technique of controlling the language has been pursued vigorously in public discussions about climate change, and especially where the three key terms global warming, climate change and greenhouse are concerned.

Global warming, climate change, greenhouse
To ask the question 'is global warming occurring?' might seem innocent enough. But the accurate answer is actually that 'it depends', and one of the things that it depends upon is what you mean by global warming in the first place (but see also Chapter 2).

Global warming is merely one of the two alternative directions of climate change. Over time periods of decades or longer, average global temperature rarely remains static but either increases or decreases in accord with natural cyclicity on many time-scales. Between about 1965-98, the instrumental record at the Earth's surface suggests that average global temperature increased by a modest few tenths of a degree, i.e. global warming *sensu stricto* occurred during the late twentieth century. However, to the general public, the phrase global warming has come to carry the meaning *'human caused* global warming', and it is simply not true that the late twentieth century temperature increase can be shown to have a primary human cause.

Between 1988 and 2005, most media reporters writing about the 'global warming' issue used that term to headline or describe it. On 3 February 2005, that changed; almost overnight, and across the world, the phenomenon became re-referenced in the public arena under the phrase 'climate change'. This redefinition, which allowed weather and climate change of all types to be beaten up as a matter of concern, did not happen by accident but was the outcome of a now infamous *'Avoiding Dangerous Climate Change'* meeting in Exeter[15], co-ordinated by the UK Meteorological Office's Hadley Centre with the close involvement of several large green NGO's. The Exeter meeting had two main aims: first, replacing the term global warming (which was no longer happening) with climate change (which always would be); and, second, adopting, for entirely political reasons, a fanciful 2°C target as the 'dangerous' amount of warming that politicians should be advised that they were to prevent. It is a tribute to the power of language, as well as depressing, to note that five years later the same two dishonesties continue to permeate nearly all public discussion of the global warming issue.

In fact, however, misuse of the term climate change significantly predates the Exeter meeting, for the term achieved the exalted status of legal misdefinition in the United Nation's Framework Convention on Climate Change (FCCC)[16]. This convention, which came into force in 1994, states in Article 1.2 that:

> 'Climate change' means a change of climate which is attributed directly or indirectly to human activity that alters the composition of the global atmosphere and which is in addition to natural climate variability observed over comparable time periods.

In FCCC diplospeak, then, 'climate change' doesn't mean climate change, but rather 'human-caused by atmospheric alteration climate change'. Humpty Dumpty comes to mind.

To add to the confusion, the IPCC, which operates under the aegis of the FCCC, uses the term climate change in a more usual scientific way, for example in its Fourth Assessment Report[5]:

> Climate change may be due to natural internal processes or external forcings, or to persistent anthropogenic changes in the composition of the atmosphere or in land use.

Finally, we should note that the term 'greenhouse' as used in the media has become a sort of shorthand for all of this, and even more. The connotations implicit in the term greenhouse include not only global warming, but also human-caused global warming, and human-caused-by-loading-the-atmosphere-with-carbon-dioxide-global-warming.

Averages

Another powerful word that is much misapplied in the climate debate is 'average', for almost the entire public discussion concentrates on perceived deleterious changes in properties such as 'global average temperature' or 'global average sea-level'.

Such averages have no physical existence, but represent instead convenient statistics that are generated from many separate pieces of data gathered from disparate places. It is precisely for this reason that there is so much argument about the accuracy of various different estimates of temperature and sea-level history. For the construction of such averages requires data to be selected, corrected and statistically manipulated, activities which may be quite legitimately undertaken in different ways by different investigators and which then may lead to different outcomes.

Real world environmental effects are not imposed by changes in global average conditions, but by changes in specific local conditions. What is of concern to the citizens of the island of Tuvalu is whether their *local relative sea-level* is going up or down (and, despite much alarmist propaganda to the contrary, it exhibits no significant long-term trend – see Chapter 4), not what an imaginary *global average sea-level* may be doing.

This point is particularly important when applied to predictions of a future, warmer world. Were the world temperature average to increase appreciably, scientific principles, computer model predictions and current trends all agree that the manifestation of this will be that much of the warming will take place at high latitudes and in winter. In some places, such warming will have virtually no environmental impact: for example, an uninhabitable ice cap at -30°C will remain an uninhabitable ice cap at -27°C. In other places, such as at the fringes of such an ice cap, warming is likely to be beneficial (from our perspective, the planet being entirely neutral about the matter) because warmer temperatures and longer growing seasons will enhance the chances of the establishment and survival of biota there.

These comments notwithstanding, it is inescapable that most of the discussion regarding climate change presented in this book has had to be framed in the same terms as the public debate, i.e. in terms of changes to 'world average temperature'. But the reader should never forget that such abstraction is far removed from the physical realities of what will happen to his or her own local environment when climate change occurs. There, the relevant questions must always be 'how are the local conditions going to change?' (with temperature and sea-level each able to go either up or down), and 'will the environmental response to that change be positive or negative?' from a human point of view (with either outcome possible at a particular location).

To simply assert, as many do, that global warming is going to take place and that its impact is everywhere going to be negative is to make neither a scientifically based nor a sensible statement. Rather, it is a statement of devotion to the green religion that has been aptly called eco-salvationism.

Coda

'*Do you believe in global warming?*' the reporter asks (meaning, of course '*do you believe in dangerous global warming caused by human carbon dioxide emissions?*').

'*It depends,*' I reply. '*For there are many different realities of climate change.*'

In chapters 1 to 4, we will turn to the first of those realities, that of science.

1 The geological context of natural climate change

At 4° C, it is cold in the storage refrigerator. One needs to rug up well to work here. I am at the US headquarters of the Ocean Drilling Programme at Texas A&M University, studying seabed cores from the southwest Pacific Ocean. As the cores pass through the logging sensor that measures their character, the rhythmic pattern of ancient climate change is displayed before me. Friendly, fossiliferous brown sands for the warm interglacial periods, and hostile, sterile grey clays for the cold glaciations....

Backwards for hundreds of thousands of years, the core alternations march. Some, metronomic in their occurrence, are ruled by changes in the Earth's orbit at periods of about 20,000, 41,000 and 100,000 years; others are paced by fluctuations in solar output on a scale of centuries or millennia; and others display irregular yet rapid oceanographic and climate shifts that are caused by we know not what. Climate, it seems, changes ceaselessly in either direction: sometimes cooling, sometimes warming, often for reasons that we do not yet fully understand.

(Bob Carter, UK *Sunday Telegraph*, 7 April 2007)

The issue of dangerous human-caused global warming is a complex one. It can be assessed meaningfully only against our knowledge of natural climate change, which is incomplete and in some regards even rudimentary. This point being understood, and as outlined below, study of the geological record of climate reveals many instances of natural changes of a speed and magnitude that would be hazardous to human life and economic well-being should they be revisited upon our planet today. Many of these changes are unpredictable, even in hindsight. That such natural changes will occur again in the future, including both episodic step events and longer term cooling and warming trends, is certain.

The geological record of climate change

The focus of the Intergovernmental Panel on Climate Change (IPCC) activity has been on erecting computer models that are rooted in a knowledge of the last 150 years of instrumented climate observations, sometimes extending back to around 1,000 years as defined by proxy measurements such as tree ring analysis. This is an utterly inadequate period over which to seek to understand climate change in the round.

Using climate records that represent the last several million years, palaeoclimatologists and palaeoceanographers have established a sound understanding of the natural patterns and some of the mechanisms of climate change. The most important evidence comes from sediment cores from beneath the deep seafloor and ice cores through the Greenland and Antarctic ice caps. Any one such core does not, of course, depict global climate; however, suitable cores yield climate data that are representative of a wide region and some may even approximate a global pattern. Generally agreed inferences from these data are as follows.

Milankovitch frequences of climatic change

Between about 6 and 3.5 million years ago, during the period that geologists term the Pliocene, small warm-cold climatic oscillations occurred around a mean temperature that was 2-3°C warmer than today (Fig. 1). Starting 3.5 million years ago, global temperature embarked on a steady decline, towards the end of which the background 41,000 year long climatic cycles became overprinted with more severe, 100,000 year long, glacial-interglacial oscillations; detailed graphs display also a 20,000 year oscillation superimposed on the other two.

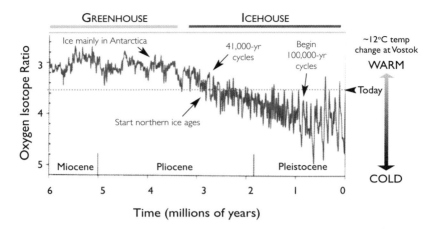

Fig. 1 Pacific Ocean deep-water temperature for the last 6 million years, showing greenhouse and icehouse phases of Earth's history. Dotted horizontal line indicates today's temperature; temperature scale bar (on right) represents the equivalent air temperature change at Vostok, Antarctica. Composite curve from ODP Sites 846 and 849, Equatorial Pacific, based upon oxygen isotope ratios measured in benthic foraminifera in marine cores.

These three fundamental frequencies of climatic oscillation are termed Milankovitch frequencies[17], after Serbian geophysicist Milutin Milankovitch who, early in the twentieth century, spent almost 20 years laboriously calculating the first handmade graphs of Earth's recent climatic history, graphs that can now be produced more accurately on a laptop computer in seconds. Milankovitch's insight, following that of the self-taught British geologist and physicist John Croll, who published the important and novel book *Climate and Time* in 1875, was to appreciate that the distribution of radiant solar energy received across planet Earth changes through time in correspondence with fluctuations in the Earth's orbit. These orbital variations are caused by gravitational interaction with the other planets of the Solar System, and affect both the tilt of the Earth's axis and the shape of its orbit around the sun (Fig. 2). Specifically, the path of the orbit varies from more to less elliptical on a 100,000 year scale; the tilt of the Earth's axis

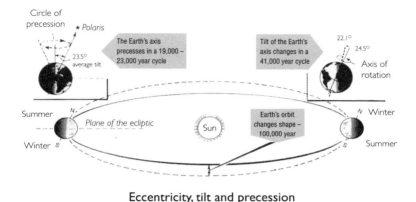

Eccentricity, tilt and precession

Fig. 2 Cartoon summary of the changes in Earth's orbital characteristics around the sun that exert control over climate, namely the 20,000 year (precession), 41,000 year (tilt) and 100,000 year (eccentricity) Milankovitch cyclicities. The changes in orbital geometry affect the seasonal distribution of sunlight across the Earth's surface, and thereby influence the growth or melting of ice, especially at high northern latitudes.

varies slightly, between about 22.1° and 24.5° on a 41,000 year cycle; and, third, the Earth's tilted axis also precesses ('wobbles') on a roughly 20,000 year cycle. The changing geometries exert a marked effect on Earth's seasonality, which in turn controls the accumulation of snow and ice at the high latitudes at which the great northern hemisphere ice sheets accumulated over the last few million years.

Since about 0.6 million years ago, after what has been termed the mid-Pleistocene (climatic) revolution, each major glacial-interglacial oscillation has occurred on the longer, 100,000-year periodicity. For more than 90 per cent of this time the Earth's mean temperature was cooler, and often much cooler (up to ~6°C), than today (Fig. 3). Warm interglacial periods comprised less than 10 per cent of the time, and on average lasted only 10,000 years. Civilization and our modern society developed during the most recent warm interglacial period (the Holocene), which has already

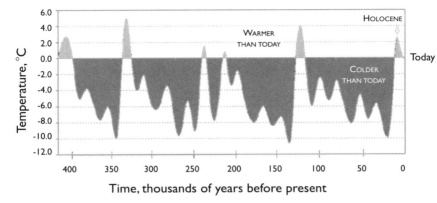

Fig. 3 Surface air temperature at Vostok Station, Antarctica, over the last 400,000 years. Horizontal zero line represents today's temperature as a reference; light grey shaded peaks above this line indicate the Holocene and previous warm interglacials, which were up to 5°C warmer than today; darker grey shaded areas below the zero line indicate past glacial periods. Curve based on measurements of deuterium isotopes (a proxy for surface air temperature) in an ice core.

lasted 10,000 years. In many places, temperatures earlier in the Holocene – during the 10-9 ka period that corresponded to a maximum in total solar insolation at 65°N[18], formerly referred to as the climatic optimum – were up 1-2°C warmer than today[19].

The spectacular Milankovitch climatic record was first captured in deep ocean sediment cores[20], and later confirmed by studies of glacial ice cores from Greenland and Antarctica[21]. Excitingly, these latter cores have the unique capability of yielding measurements of ancient atmospheric chemistry from air samples that are captured as bubbles within the ice. Measurements on the gas contained in the bubbles demonstrate that changes in past temperature and atmospheric carbon dioxide concentration in such cores occur in close parallelism. In detail, however, the changes in temperature *precede* their parallel changes in carbon dioxide by between ~800 and 2,000 years[22]. This vital point establishes that carbon dioxide cannot be the primary forcing agent for temperature change at the glacial-interglacial scale. Strangely, this obvious conclusion continues to elude Al Gore and other alarmist commentators on the global warming issue.

Abrupt climatic events

The recent climate record is not only cyclic; it is also punctuated by episodes of abrupt climate change, when climate sometimes traversed almost its full glacial-interglacial range in a period as short as a few years to a few decades. A well-known and dramatic example of this concerns a post-last-glacial warming in the northern hemisphere, where Greenland ice cores and other records show that abrupt warming to almost full interglacial level 14,500 years ago was followed by a reversion to temperatures of glacial intensity termed the Younger Dryas episode, which in turn was followed by a sharp resumption of warming 11,600 years ago that continued more gently up to the start of the Holocene (Fig. 4). The two episodes of sharp warming occurred in three years and 60 years respectively, whilst the cooling into the Younger Dryas stretched over 1,500 years, after which its deepest cold phase continued for another 1,000 years[23]. The causes of such abrupt climate changes as these remain largely unknown – though changes in ocean currents, changes in the delivery of fresh water to the ocean (including by the sudden drainage of giant glacial lakes), changes in ocean current flow and even extraterrestrial impact remain among the candidate explanations.

Younger Dryas summers and winters are estimated to have been 4°C and 28°C cooler than today's, respectively. A computer model reconstruction of sea ice extent during Younger Dryas winters projects that the ice advanced to cover nearly all of the North Atlantic Ocean north of a latitude of 40°N[24], which corresponds approximately to a line connecting the Brittany Peninsula (France) and northern Newfoundland (Canada). Apparently, very little commercial shipping would have been coming out of Europe's Channel ports. Noting that ice caps take many thousand years to grow or melt, but that sea ice can be built in a year or two, one wonders idly what contingency plans have been laid by north Atlantic nations for the recurrence of a sharp climatic cooling of the Younger Dryas type – with attendant dislocation of much transatlantic marine commerce by sea ice, and an increased demand for energy for heating due to the intense cold. The last time that I looked, however, European and US governments seemed to be concerned only with *l'affaire du jour* of global warming.

Temperatures during the Holocene

Twenty thousand years ago, the peak of the last great glaciation, is but a geological heartbeat away. Two hominid species inhabited our planet during the earlier parts of this ice age, *Homo neanderthalensis* and *Homo sapiens*; life must have been tough, caves very welcome and the ability to light fires an absolute blessing. The climatic warming that then ensued was punctuated by the Younger Dryas, and ended 10,000 years ago at the dawn of the Holocene. Starting around 12,000 years ago, during the cultural period called the Mesolithic, *Homo sapiens* discovered how to make pottery, farm crops and domesticate animals, how to smelt first bronze and then iron, and how to develop city civilizations – many of these developments surely being aided by the relative warmth of the climate.

Compared with the vicissitudes of the Younger Dryas, the Holocene climate may appear to have been relatively stable, but nonetheless, and as always, change was a constant. First, the long-term temperature record of the Holocene is one of cooling by 1-2°C from the post-glacial climatic optimum in the early Holocene (Fig. 4). Second, throughout the Holocene, a 1,500-year long climate cycle, similarly of 1-2°C magnitude and sometimes called the Bond Cycle, was conspicuous[25]. In Greenland, the three most recent historic warm peaks of the Bond cycle (Minoan, Roman and Mediaeval Warm Periods) all attained or exceeded the magnitude of the late twentieth century warming (Fig. 5). A variety of detailed proxies for past temperature show a similar climatic pattern in many localities around the world[26], and some of the best of these

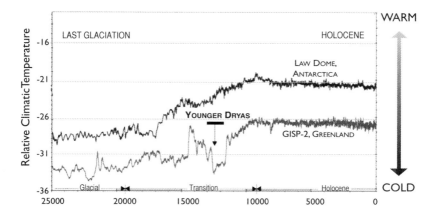

Fig. 4   Surface air temperatures since the Last Glaciation (20,000 years BP) compared for Antarctica (Law Dome ice core) and Greenland (GISP-2 ice core), based on ice core measurements of deuterium and oxygen isotopes, respectively. Note the conspicuous climatic cooling event called the Younger Dryas in Greenland, and that the last 10,000 years (the Holocene) comprises a period of gentle cooling that continues today.

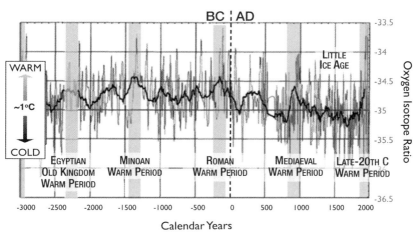

Fig. 5   Greenland surface air temperature for the last 5,000 years (black line, moving average). Short warm periods like the Late Twentieth Century Warming occur about every 1,500 years (grey shading), separated by longer cool periods such as the Little Ice Age. Curve based on measurements of oxygen isotopes from the GISP-2 ice core, Greenland.

Corrected Global Temperature Reconstruction

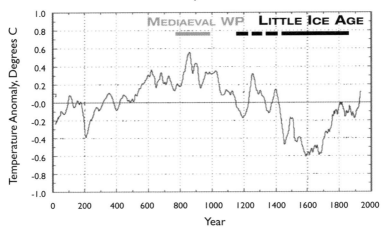

Fig. 6 Global temperature for the last 2,000 years, reconstructed from 18 sets of geological proxy measurements, including pollen and diatoms from lake cores and isotope data from ice cores and speleothems. Note the clear delineation of the Little Ice Age and Mediaeval Warm Period, and that the MWP was significantly warmer than the twentieth century up until 1980.

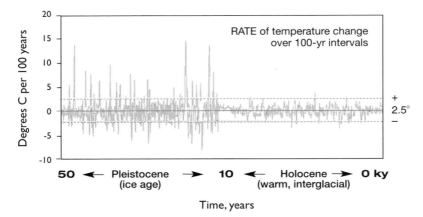

Fig. 7 Rate of temperature change over the Greenland ice cap in degrees/century for the last 50,000 years. Note that the rate of change during the Holocene fluctuated between + and − 2.5°/century of both warmings and coolings (between the two horizontal dotted grey lines). This rate of change compares with a twentieth century warming rate of ~1°/century for Greenland. Based on oxygen isotope measurements in the GISP2 ice core.

records have been used to construct a global temperature estimate for the last 1,500 years that confirms the greater warmth of the Mediaeval over the Late Twentieth Century Warm Period (Fig. 6)[27].

Is modern climate change unusual?
Importantly, compared with the ancient climate record, temperatures during the late twentieth century were neither particularly high nor particularly fast-changing. Not only were recent Bond Cycle peaks a degree or so warmer than modern temperature, but temperatures in Antarctica for the three interglacial periods that preceded the Holocene were up to 5°C warmer than today (Fig. 3), and temperatures ~2-3°C warmer probably characterized much of the planet during the Pliocene (Fig. 1). Meanwhile, as to the rate of temperature change, analysis shows that the two twentieth century warming pulses occurred at a rate of ~1.5°C/century, which falls well within the natural rates of Holocene warming and cooling exhibited by high quality records like the Greenland ice core (Greenland Ice Core Project; Fig. 7).

It is clear from these various facts that a warmer or cooler planet than today's is far from unusual, and that nature recognizes nothing 'ideal' about mid-twentieth temperatures. It is also clear that climate changes naturally all the time. The idea that is implicit in much public discussion of the global warming issue – that climate was stable (or constant) prior to the industrial revolution, after which human emissions have rendered it unstable – is fanciful. Change is simply what climate does.

Climatic cycles of solar origin
The Earth's surficial skin, within which we live and travel, has two main sources of heat energy. Some heat comes from deep within the planet, derived by conduction from the molten core and the decay of radioactive elements in the mantle and crust, in the very small amounts of 20-40 milliwatts/square metre (mW/m^2), depending upon whether it is measured above oceanic (hotter) or

continental (cooler) crust[28]. But the dominant energy source by far is the sun, whose direct radiation provides ~342 watts/square metre (W/m^2) of heat at the top of the atmosphere. It is this solar heat, or more strictly its redistribution by radiation and convection, that drives the climate system.

It is long known that the sun's direct radiation of light and heat (which scientists term the total solar irradiance – TSI) varies slightly on the scale of the ~11 year sunspot cycle, during which the sun builds up from a small to a large number of spots, which then decline in number again (Fig. 8a,b). Superimposed on top of the sunspot cyclicity during historic times has been a long-term trend of slight increase in overall irradiance, which amounts to ~2 W/m^2 since the seventeenth century (Fig. 8a). Supporters of the IPCC, noting that this increase represents only 0.1 per cent of the total solar irradiance, are fond of claiming that this small change is inadequate to explain the increases observed in global temperature since the nineteenth century. But the main way in which the changes in the sun's irradiance affect Earth's climate is through their direct influence on seasonal and annual climate cycles, which is a much more important effect than the minor and indirect influence of increasing atmospheric carbon dioxide. Also, to restrict one's attention to changes in irradiance only – which the IPCC does in arguing that because the sun exhibits only minor long-term irradiance change it can have but little effect on climate – is to throw the baby out with the bathwater in spectacular fashion, for in reality there are many other important ways in which solar variations affect Earth's climate.

The other aspects of Earth-sun energy interrelations that are known to play a role in climate include:

- Variations in the intensity of the Sun's magnetic fields on cycles that include the Schwabe (eleven year)[29], Hale (22 year)[30] and Gleissberg (70- 90 year)[31] periodicities.
- The effect of the sun's plasma and electromagnetic fields on rates of Earth rotation, and therefore the length of day (LOD)[32].

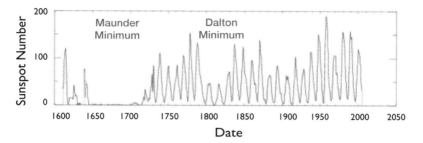

Fig. 8a The observational record of sunspot cycles since the early seventeenth Century. Note (i) the rhythmic 11-year periodicity; (ii) an overall gradual increase in average solar intensity across the length of the record; and (iii) the occurrence of episodic periods of low (Dalton Minimum) or absent (Maunder Minimum) sunspot activity, which are associated with strong planetary cooling.

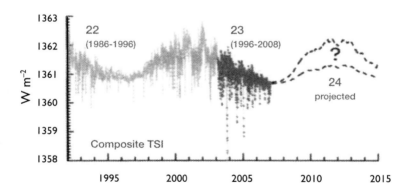

Fig. 8b Solar radiation (TSI) received at the top of the atmosphere, measured from satellites for Solar Cycles 22-23, with two alternative intensities projected for Cycle 24. Given that the cycle 23-24 solar minimum lasted 3 years longer than average, the lower of the projected Cycle 24 trajectories is the more likely, and will be accompanied by global cooling.

- The effect of the sun's gravitational field through the 18.6-year long Lunar Nodal Cycle – which causes variations in atmospheric pressure, temperature, rainfall, sea-level and ocean temperature, especially at high latitudes[33].

- The known links between solar activity and monsoonal activity[34], or the phases of climate oscillations such as the Atlantic Multi-decadal Oscillation, a 60 year-long cycle during which sea surface temperature varies ~0.2°C above and below the long-term average, with concomitant effects on northern hemisphere air temperature, rainfall and drought.

- Magnetic fields associated with solar flares, which modulate galactic cosmic ray input into the Earth's atmosphere, and in turn may cause variations in the nucleation of low-level clouds at up to a few km height. This causes cooling, a one per cent variation in low cloud cover producing a similar change in forcing[35] (~4 W/m^2) as the estimated increase caused by human greenhouse gases. This possible mechanism is controversial and remains under test in current experiments devised by Henrik Svensmark[36] at the European Organization for Nuclear Research (CERN). But irrespective of the results of these experimental tests, and of the precise causal mechanism, Neff *et al.* (2001)[37] have provided incontrovertible evidence from palaeoclimate records for a link between varying cosmic radiation and climate. Using samples from a speleothem from a cave in Oman, Middle East, these authors showed that a close correlation exists between radiocarbon production rates (driven by incoming cosmic radiation, which is solar modulated) and rainfall (as reflected in the geochemical signature of oxygen isotopes) (Fig. 9).

- As already discussed under *Holocene* above, the 1,500 year-long Bond Cycle is probably of solar origin. Another climate rhythm of similar length occurs in glacial sediments deposited about 90,000-15,000 years ago, especially in the North Atlantic region[38]. Called a Dansgaard-Oeschger, or D-O, event, this cycle may also be a response to solar forcing.

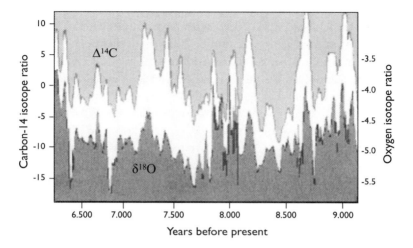

Fig. 9 Comparison of proxy records from Oman for the amounts of incoming cosmic radiation (proxy: radioactive ¹⁴C) and rainfall (proxy: oxygen isotope signature) for a 3,500 year long period during the early Holocene, measured on a speleothem from Oman. The close correlation between these two curves indicates the existence of a link between cosmic rays and climate, perhaps through low cloud formation; see text for more explanation.

The IPCC represents this large corpus of research into Sun-Earth climate relations (and other non-greenhouse gas causes of climate change such as 1,800 and 5,000 year lunar tidal cycles[39]) poorly in its assessment reports, and nor has any significant attempt been made to include such solar effects in the IPCC's modelling. As a result, the current generation of general circulation models (GCM) underestimate the variability and changes from solar-induced climate signals. That many of the mechanisms and possible mechanisms by which the sun influences Earth's climate are poorly understood is no justification for ignoring them.

The transition between solar sunspot cycles 23 and 24
The average length of solar cycles, from minimum to minimum, is eleven years. The solar minimum between cycles 23 and 24

occurred in December 2008[40], making cycle 23, at 12.5 years long, the longest since 1823. However, the sun remained in a quiescent state through most of 2009, with only intermittent cycle 24 sunspots occurring; by the end of 2009 there had been 771 days without sunspots during the transition.

It is established from observation that solar cycles longer than the eleven year average are followed by later cycles of lesser intensity, and, commensurately, a cooler climate[41]. Solar Cycle 23 was three years longer than Cycle 22. Based on the theory originally proposed by Friis-Christensen and Lassen, this implies that cooling of up to 2.2°C may occur during Cycle 24 (compared with temperatures during Cycle 23) for the mid-latitude grain-growing areas of the northern hemisphere[42].

The Average Planetary Magnetic Index (AP index) is a commonly used yardstick to indicate the Earth's magnetic condition and its close relation to solar magnetic variability. Historically, this index recorded a lowest monthly value of 4 in January 1932 – until recently. Between November 2008 and September 2009, the AP index returned persistent readings of 4's and 5's, sinking to 3's in October and November, and finally to 1 in December – which is the lowest reading in the 165 years of observations since 1844[43]. At the same time that the sun's magnetic field was setting new record lows, the northern hemisphere winter turned into one of the most severe in decades. It was therefore not surprising that during 2009 NASA's chief solar scientist, Dr David Hathaway, commented that 'something like the Dalton Minimum — two solar cycles in the early 1800s that peaked at about an average of 50 sunspots — lies in the realm of the possible'[44]. Given the implications of the recurrence of the cold conditions that characterized the Dalton Minimum, it is astonishing that the IPCC, and the governments that it advises, continue to ignore the implications of the cycle 23/24 solar transition, and the now probable outcome of significant cooling and decrease of crop yields over the next few decades.

One final point about solar influence

Most of the assertions that solar causes can only play a minor role in controlling the late twentieth century warming are misplaced, because their authors not only fail to consider all the mechanisms listed above, but also fail to allow for the lags that occur in the redistribution of solar heat through various planetary heat transfers and storages. As US astrophysicist Willie Soon has shown,[45] delays of 5-20 years in the manifestation of a solar signal are associated with ocean storage and circulation of heat. The peak solar radiation outputs observed in the 1980s and early 1990s, with weakening since then, match well with climatic observations when applied with an appropriate time lag.

In summary, the argument advanced by the IPCC – that the sun can only affect climate through irradiance, that irradiance changes are small, and so the sun cannot play a major role in global warming or cooling trends – is incorrect.

The human influence on climate change in context

Despite the great variability and high magnitudes of natural climate change, it is clearly also the case that human activities have a measurable effect on local climates.

For example, the concrete, glass, steel and macadam that are used to build a conurbation absorb more radiant heat from the sun during the day than did the pre-existing natural vegetation. The result is a local warming called the urban heat island effect, which, for a large city, has a magnitude of several degrees[46].

Alternatively, when humans clear forested areas, the pasture or crops that are planted are often lighter in colour than was the forest. This results in reflection of more of the incoming solar energy than before, and hence cooling[47].

So humans, through changed land usage, have an effect on local climate that is variously warming or cooling. Summing these local signals all over the globe, it follows that humans must exercise an effect on global climate also. The question in context, therefore, is not 'do humans have an effect on global climate?' but rather, 'what is the sign and magnitude of the net human effect on global climate?' As you will see, no-one knows the answer.

Local, regional and global climate changes naturally all the time, and human activity is definitely known to cause local change. Yet remarkably, given the expenditure and effort spent looking for it since 1990, no summed human effect on global climate has ever been identified or measured. Therefore, the human signal most probably lies buried in the variability and noise of the natural climate system. This is so to such a degree that as a statement of fact we cannot even be certain whether the net human signal is one of global warming or cooling[48]. Though it is true that many scientists anticipate on theoretical grounds that net warming is the more likely, no direct evidence exists that any such warming would *ipso facto* be dangerous.

We therefore turn next to a discussion of the evidence for climate change, both warmings and coolings, that is contained within the meteorological record.

2 Dangerous Twentieth-Century warming? The short meteorological record

The instrumental record of climate change:
150 years of thermometer data = 5 climate data points
50 years of radiosonde data = 1.6 climate data points
30 years of satellite data = 1 climate data point

Three primary arguments are asserted in support of the idea of dangerous, modern global warming. The first is that computer models '*predict*' it will be so, which idea is discussed in Chapter 5. The second is that a variety of other real-world phenomena (ice volume, ocean temperature, polar bear numbers and so on *ad infinitum*[49]) are changing in a way that is consistent with warming, the topic of Chapter 6. And the third, discussed in this chapter, is that twentieth century temperature *measurements* demonstrate that dangerous climatic warming is occurring.

We have seen in Chapter 1 that climate change needs to be studied in the context of geological time-scales, and that these time-scales range all the way up to millions of years. Such conclusions, though, are drawn from proxy, geological data, and need to be considered alongside the evidence from actual measurements of temperature. The mercury thermometer was invented in the early eighteenth century, but estimating a global temperature requires accurate observational records from a suitable worldwide network of reliable thermometer observing stations.

Most meteorologists agree that an adequate network of observations is only available back to about 1860. In essence, then, meteorologists have only a 150 year-long record on which to ply their craft, and a question that is seldom addressed is to what degree this weather record might also be a climate record. Self-evidently, the instrumental record and related research into the physics and chemistry of the atmosphere are paramount for what they teach us about the *mechanisms* of weather (and hence climate) change, but it is far from self-evident that the meteorological record tells us much about the *history* of climate. This chapter, then, addresses these questions, to help enable the reader to judge what the short meteorological record can really tell us about climate change. But first, the climate normal.

Weather, and the climate normal

Mark Twain reputedly once remarked 'Climate is what you expect; weather is what you get'. Though it is hard to improve on such a pithy aphorism, scientists prefer to use definitions that are quantitatively based. Accordingly, since the International Meteorological Organization's (IMO) 1935 conference in Warsaw it has been agreed among climatologists that 'climate' is taken to be represented at a particular site by an averaged 30 year-long span of meteorological data, called the *climate normal*. The period 1901-30 was nominated by the IMO as the first such normal, to be supplanted later by 1931-60 and the current 1961-90 periods. The World Meteorological Organization (WMO, successor to the IMO) only undertakes a comprehensive analysis for a new climate normal every 30 years, but some individual countries, including the USA, update their normal period every decade, e.g. 1971-2000 and, shortly, 1981-2010.

Though some justification exists for making such regular shifts in the climate reference period, the practice can cause confusion, especially among non-scientists. It also opens the way to bias in the presentation of temperature data, whereby a particular climate normal period is preferred as a reference because using it delivers an outcome desired by the researcher – such as apparently enhanced warmth or cold. Certainly, shifting the climate normal from time to time doesn't assist making accurate judgements about climate change, which requires the analysis of long-term climate proxy records that are plotted against an unchanging baseline.

The use of climate normals also opens the way for scientific method to be traduced by convolving it with sociological or political practices. As social scientists Hulme *et al.* remarked in 2009[50]:

> ... the choice of these statistical 'normals' reflects cultural, political and psychological preferences and practices as much as scientific ones...Expectations of the climatic future are influenced by social as well as statistical norms. Seeing climate as co-constructed between the psychocultural constraints of society and the physical constraints of the material world offers a different way of thinking about the instabilities of climate and the ways we adapt to them.

Quite so. For it is precisely this 'different' way of thinking that has led us into the confused morasse of scientific, sociological and political advice that is delivered by the Intergovernmental Panel on Climate Change (IPCC).

Whilst acknowledging their limitations, the recognition of climate normals is nonetheless useful in various ways, one of which is to allow the direct comparison of temperature graphs from different climatic zones around the world. This is accomplished by plotting temperature measurements as departures from the 30-year average as 'anomalies' that differ from the adopted 'climate normal' (Fig. 10), rather than as raw temperature measurements, which are obviously difficult to compare directly between particular stations because the actual magnitude of daily temperatures varies widely with latitude and other factors.

Our longest weather records
The primitive temperature measuring instruments termed thermoscopes, forebears of the modern thermometer, were

Departures in temperature in deg. C from the 1961–1990 average

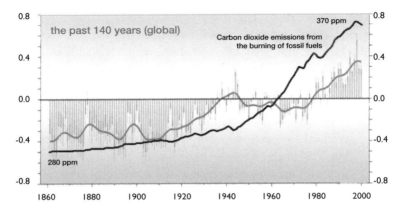

Fig. 10 Standard atmospheric surface temperature since 1860, determined by averaging thermometer measurements worldwide (UK Hadley Centre). Superimposed black line represents the graph of human carbon dioxide emissions. Note that global temperature cooled for 30 years at the very time (1940-1970) that emissions were rising most rapidly.

invented in the late sixteenth to early seventeenth centuries by Italian scientists Santorio Santorio and Galileo Galilei. Other scientists quickly grasped that this made possible the collection of useful meteorological measurements. The longest established ground temperature record, the first part of which was captured by these instruments, is termed the Central England Temperature (CET) index and starts in 1659[51] (data and plotted graph from 1772, available though the UK Meteorological Office[52]). The CET index has been analysed recently by Scottish chemist Wilson Flood[53]. Though the 351 year-long record demonstrates an overall warming from the last Maunder Minimum cooling of the 1690s, interestingly 'the average CET summer temperature in the eighteenth century was 15.46°C while that for the twentieth century was 15.35°C. Far from being warmer due to assumed global warming, comparison of actual temperature data shows that UK summers in the twentieth century were cooler than those of two centuries previously.'

However, to create an estimate of changing *global* temperature obviously requires the existence of a number of such measuring stations located around the globe, which took another 200 years to accomplish. This is why the graph deployed by the IPCC plots the global temperature statistic only from 1860 onward[5] (Fig. 10). Consequently our longest instrumental estimate of global temperature (at 150 years) comprises just five climate data points, and our longest single meteorological temperature record (CET, at 351 years) just over twice that number. This paucity of the instrumental 'climate' record is worth bearing in mind next time that someone tries to convince you that we should revolutionize the energy systems of our industrial societies on account of dangerous human-caused global warming.

Recently, new methods of measuring atmospheric temperature have been developed using thermistors mounted on weather balloons and measurements of the brightness of oxygen emissions (which relate directly to temperature) with satellite-mounted instruments. These temperature records, whilst highly accurate, are available only since 1958 (nearly two climate data points) and 1979 (one climate data point), respectively.

Despite it being more a weather than a climate record, a great deal of valuable information resides in the instrumental dataset, especially with regard to helping us understand meteorological processes. The two main ground thermometer temperature compilations made from this data by the British Meteorological Office[54] and NASA (Goddard Institute for Space Studies)[55] are similar, and show an overall rise in temperature of a little less than 1°C since 1860, this warming in part representing recovery from the earlier Little Ice Age (Fig. 6)[56]. Warming, however, did not proceed in a single, steadily rising curve. Instead, and as for all extended climate records of adequate resolution, the thermometer data display a multi-decadal rhythmicity with alternating periods of warming and cooling. Within this record, it is the short phase of mild warming that started around 1979 and terminated in 1998 that so excites the IPCC and other climate alarmists.

Though it is in widespread use, the thermometer temperature dataset is far from perfect; for its earlier part is based on rather too few high quality station records, and its later part, since around 1980, is known to be contaminated by the urban heat island effect and by the closure of many rural stations in the 1990s[57]. This deficiency has been underlined by information that the Climategate scandal, and related questioning, has pitched into the public domain (Chapter 12). For example, a previously undistributed Climatic Research Unit (CRU) report with an error estimate for 1969 temperatures shows widespread, worldwide temperature errors of 1-5°C[58], i.e. errors that far exceed the claimed late twentieth century warming. In effect, no statistically significant warming will be able to be inferred until the current temperature rises well over 1°C above that computed for 1969.

Compounding the situation even more, it has now become all too evident that the datasets that underpin the official thermometer record in all its guises have been subject to heavy manipulation in order to 'correct' them. For countries as widely scattered as USA, UK, NZ, and Australia[59], temperature measurements from a sample of rural stations reveal no statistically significant warming during the twentieth century, as opposed to the warming identified in the government sanctioned, 'corrected' temperature results. US meteorologists Joseph D'Aleo and Anthony Watts, after analysing the corrections applied to global temperature datasets, have recently come to the remarkable conclusion that

> Instrumental temperature data for the pre-satellite era (1850-1980) have been so widely, systematically, and undirectionally tampered with that it cannot be credibly asserted there has been any significant 'global warming' in the twentieth century[60].

A comparison between the inadequate thermometer datasets and two other independent and more accurate datasets is salutory. The first of these, collected using radiosonde sensors mounted on

WEATHER BALLOON

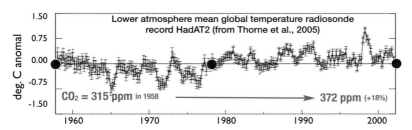

Fig. 11a   Estimated lower atmosphere global temperature record since 1958, based on measurements from weather balloons. Note the presence of (i) cooling from 1958 to 1977; (ii) warming, mostly as a step in 1977, from 1977-2005; and (iii) no net warming between 1958 and 2005. Over the same time period there has been an 18% increase in atmospheric carbon dioxide. Black dots denote times at which the temperature falls upon the zero anomaly line, i.e. no net change has occurred between them.

SATELLITE

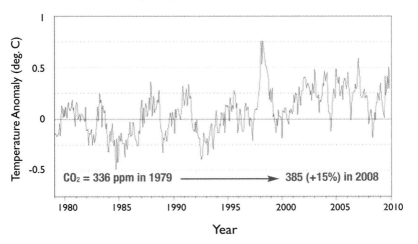

Fig. 11b   Estimated global lower atmosphere temperature since 1979, based on measurements from satellites. Note (i) that for the period of overlap, the record matches that of the weather balloons (Fig. 11a); and (ii) that the small amount of warming between 1979 and 2008 is again not represented by a linear trend but by a single step across the 1998 El Nino warm spike.

weather balloons, indicates a cooling between 1958-75, followed by an equivalent warming to 2002, i.e. displays no significant overall warming between 1958 and 2002 (Fig. 11a)[61]. The second, collected since 1979, is compiled independently by Remote Sensing Systems (RSS) and the University of Alabama, Huntsville (UAH) from measurements made with microwave sensing units (MSU) mounted on orbiting satellites[62]. Both versions of the MSU data show the same phase of unexceptional late twentieth century warming that is exhibited by the thermometer and radiosonde records since 1979 (Fig. 11b). The warming 'trend' displayed by both these datasets can alternatively, and probably preferably, be represented as a single step increase of ~0.2°C across the 1998 El Nino event[63].

Importantly, the gentle warming that occurred (perhaps only as a 1998 El Nino step) in the late twentieth century falls well within previous natural rates and magnitudes of warming and cooling. It is therefore *prima facie* unalarming, especially in light of the comments made above about the urban heat island effect (UHI) warm bias that is present in the ground temperature records cited by the IPCC. However, irrespective of the way in which the 1979-2005 data are interpreted, it remains the case that the late twentieth century phase of rising temperature terminated in 1998. Using the MSU satellite data, Loehle (2009)[64] has shown that a statistically significant cooling has occurred over the last 12 years (cf. Fig. 12) despite an increase in atmospheric carbon dioxide of ~15 parts-per-million (ppm) or 5 per cent.

To attempt to assess the dangers of climate change on the basis of arbitrary lines fitted through temperature series that represent 1, 2, 5 or even 11 climate data points, as the IPCC largely does, is clearly a futile exercise. By overemphasizing the trivially short instrumental record, and greatly underemphasizing the varied changes that exist in geological records, which comprise many thousands of climate data points, the IPCC signals its failure to comprehend that climate change is as much a geological phenomenon as it is a meteorological one.

UAH MSU and Hadley Monthly Temperatures

Global average temperature from ground thermometer (HadCRUT3)
and lower atmosphere (UAH MSU) measurements, 2002-2009

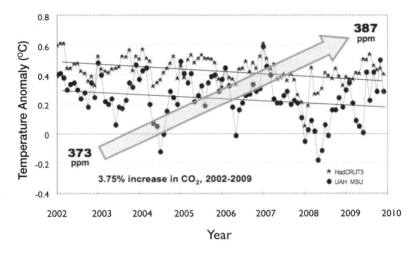

Fig. 12 Global lower atmosphere temperature measured from satellites and ground thermometers, showing cooling since 2002. Over this period, atmospheric carbon dioxide has increased by 3.75%, yet no warming has occurred.

Is there really a difference between weather and climate?

In August 2005, the New Orleans landfall of Hurricane Katrina gave Americans a sharp reminder of the danger of natural climatic events. For despite acres of newsprint to the contrary, no substantive evidence has emerged that Katrina was other than a natural storm. And, in 2009, it was Australia's turn. In early February, and, given the government's then intended carbon dioxide taxation timetable, bang on cue to enlighten Australian politicians about the difference between real and imaginary climatic events, Mother Nature provided us with another reminder of the awesome power that lurks within her natural furies.

In north Queensland, monsoonal downpours of more than 1.2 metres in seven days caused large parts of tropical Australia to be flooded; rivers rose up to 12 metres above their normal level, and 62 per cent (580,000 square kilometres) of Queensland was designated as a disaster area. At the very same time, large areas of the southern state of Victoria were ravaged by bush firestorms that razed an area greater than 4,500 square kilometres, in the process destroying more than 2,000 houses, leaving more than 7,000 people homeless and causing more than 170 deaths.

Hurricane Katrina and these Australian disasters were natural events, certainly, but were they related to weather or climate? As already outlined, meteorologists distinguish between the two concepts by characterizing weather as daily, seasonal and year on year variations at a particular place, and climate by using a 30 year-long period of, say, daily average temperature. To say that this semantic procedure is a statistical convenience (which it is) rather than a scientific truth, is to explain it rather than to criticize it. For Mother Nature does not distinguish between weather and climate in this fashion. Rather, her daily weather vagaries are driven by the same processes as her longer-term climate moods. There is no observable change in character of weather/climate records at the magic mark of 30 years, and nor is there any meaningful theoretical (as opposed to arbitrary statistical) basis for drawing such a distinction[65].

In reality climate processes – which are dynamic, non-linear, and a manifestation of heat transfer and distribution throughout two interconnected, turbulent, fluid envelopes (the ocean and the atmosphere) – occur over all time-scales from seconds up to millions of years in length. For example, the formation of new ice crystals as the nuclei of clouds takes place in microseconds. At the other end of the scale, the slow drifting of continents results in the creation of ocean basins within which currents act as a major heat distribution agent, such processes take place over millions and tens of millions of years. And these are but two of the myriads of physical, chemical and biological processes that have an influence on shaping both weather and climate. As summarized by US National Center for Atmospheric Research (NCAR) modeller James Hurrell and colleagues recently[66]:

The global coupled atmosphere-ocean-land-cryosphere system exhibits a wide range of physical and dynamical phenomena with associated physical, biological and chemical feedbacks that collectively result in a continuum of temporal and spatial variability. The traditional boundaries between weather and climate are, therefore, somewhat artificial.

In this context, then, Hurricane Katrina, the two more recent natural disasters in Australia, and myriads of other similar events around the world, are not 'just weather events' and nor are they in any way unusual. Rather, they represent typical climatological hazards from among the wide spectrum of such events that planet Earth has ever been heir to.

Is it warming or cooling: how long is a piece of string?

Another basic question that is seldom posed in public is: 'How would we recognize dangerous global warming were it to occur, and is it occurring now?' This question remains unasked despite the fact that dangerous change caused by human activity is widely assumed by many persons and organizations to have been occurring since the late 1970s. One obvious answer is 'By observing a peak of temperature, or a rate of increase, that greatly exceeds previous natural limits'. We have seen earlier (Chapter 1, and in the section above) that no such observations exist.

Nevertheless, enthusiastic, pencil-wielding journalists often ask me: 'Well, Dr. Carter, is global warming happening then?' Despite its apparent innocence, there is no simple or unambiguous answer to this question, for it is equivalent to asking 'How long is a piece of climate string?' In a little cited paper that is now almost ten years old, but the message of which is timeless, American geologists Davis and Bohling[67] showed that the only possible answer to the reporter's question, of a style that is never welcome to the press, is 'it depends.' Let me explain.

The graph in Fig. 13 displays measurement of the changing oxygen isotope ratio in a Greenland ice core over the last 17,000

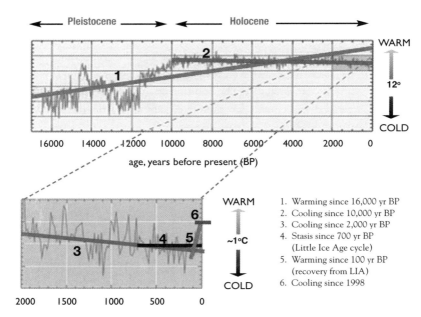

Fig. 13 Surface atmospheric air temperatures over Greenland for the last 17,000 years, based on oxygen isotope measurements in the GISP-2 ice core. Depending upon the length of time considered, the temperature trend can be viewed as increasing (since 17,000 yr bp), decreasing (since 10,000 and 2,000 year BP), static (since 700 yr BP), increasing (since 1900) or decreasing (since 1998). See text for full explanation.

years, which is a known proxy for local air temperature. Inspection shows that warming has taken place since 17,000 years ago, and also since 100 years ago. Over intermediate time periods, however, cooling has occurred since 10,000 and 2,000 years ago, temperature stasis characterizes the last 700 years and (globally, from meteorological records) a slight cooling the last ten years. Considering these facts, is the temperature in Greenland warming or cooling?

Both the ten and 100 year-long intervals of temperature change are in fact too short to carry statistical significance regarding long-term climate change. But the last 100 years of the

Greenland temperature record is still important, despite representing only about three climate data points. Because, corresponding as it does to roughly the time span of global instrumental temperature records, it confirms a twentieth century warming for which both the rate of change and the magnitude fell well within previous natural changes experienced in Greenland; and the same characteristics apply to temperature records from many other localities around the world.

Only one general conclusion is possible from this discussion. It is that the slope and magnitude of temperature trends inferred from time-series data – and therefore the answer to the question as to whether it is warming or cooling – depends entirely upon the choice of data end points. In more vernacular terms, every answer to the question that our reporter posed is underlain by cherry-picked data.

14 out of the last 15 years are the hottest 'ever'

Australia's climate minister, Penny Wong, following a well-trodden path, wrote recently that: 'Globally, 14 of the 15 warmest years on record occurred between 1995 and 2009', and reported that a Bureau of Meteorology report had recently determined '2009 was the second hottest year in Australia on record and ended our hottest decade. In Australia, each decade since the 1940s has been warmer than the last.'[68]

Variations on these statements have been endlessly repeated by climate alarmists around the world. As mantras for sloganeering they are deadly effective, as was intended by the focus group that first devised the idea, and they draw their power from the deliberate misuse of the key phrase 'on record', which in context means the trivially short instrumental record of meteorological measurements.

The suggestion that meaningful judgements about climate change can be made on the basis of the instrumental records alone (i.e. about 150 years of record, which, remember, represents five climate data points) is, putting it politely, disingenuous. When a climate record of adequate length is examined – i.e. one at least tens of thousands of years in length – and multi-decadal cycling is taken into account also, it becomes obvious that it is predictable rather than surprising that warm years are clustered around the end of the twentieth century.

Warming through the late nineteenth and twentieth centuries represents a recovery from the last severe stages of the Little Ice Age, superimposed on which are natural multi-decadal climate cycles[56]. The recovery from the Little Ice Age to a late twentieth century peak corresponds to the latter part of the present millennial climate cycle. Like its Mediaeval Warm Period predecessor, this peak will be followed by cooling, which may indeed have already started. At the same time, temperatures at the end of the twentieth century were not unusually warm. For example, they were about a degree cooler than obtained during the Holocene climatic optimum (several thousand years ago), about 2°C cooler than obtained during the last interglacial period (125 thousand years ago), and about 2-3°C cooler than obtained during the Pliocene (6-3 million years ago) (Figs. 1, 3-5). It therefore remains possible that we have yet to attain the maximum warmth of the millennial climate peak of modern times.

Like weather, climate changes episodically and often cyclically, and it is simply absurd to pick a period around the end of the twentieth century that lies near the top of a known millennial cycle, compare it with the short, incomplete 150-year record of instrumental observations, and draw a portentous inference about climate change.

Minister Wong's statement, and others like it, are scientifically trivial and appear to be deliberately intended to mislead. In reality, it is no more significant that 14 of the last 15 years are the warmest since instrumental records began than it is that the hottest days of each year cluster around and shortly after midsummer's day.

Considering also the geological context provided in Chapter 1, we have shown in this chapter that 150 years of meteorological measurement provides an inadequate data series to provide much useful information about real climate change. We have also seen that short-term warming or cooling trends characterize nearly all weather and climate time series, and that the latest such trend, since 1998, is one of cooling.

The practice, promulgated by the IPCC, of endlessly analysing short trend lines fitted in carefully selected ways through temperature data that is inherently cyclic has nothing to do with science and everything to do with politics. It is a dubious exercise that lacks intellectual rigour, and especially so when only short periods of time are under consideration. The seemingly endless press discussions about the minutiae of modern year-to-year or decade-to-decade changes in temperature relate more to angels, pins and politics than they do to scientific hypothesis testing.

3   Climate sensitivity – how now CO$_2$?

What historians will definitely wonder about in future
centuries is how deeply flawed logic, obscured by shrewd
and unrelenting propaganda, actually enabled a coalition of
powerful special interests to convince nearly everyone in
the world that carbon dioxide from human industry was a
dangerous, planet-destroying toxin.

It will be remembered as the greatest mass delusion in the
history of the world – that carbon dioxide, the life of plants,
was considered for a time to be a deadly poison.

(Ed Ring, 2008)[69]

The Kyoto Protocol recognizes six greenhouse gases that result from
human activity (Table 1). Despite the more powerful warming
effects of the other five types of gas, it is overwhelmingly carbon
dioxide that receives the greatest public and political attention.

Carbon dioxide	(1)
Methane	(21)
Nitrous oxide	(310)
Hydrofluorocarbons	(13 species: 140-11,700)
Perfluorocarbons	(7 species: 6,500-9,200)
Sulfur hexafluoride	(23,900).

TABLE 1. Greenhouse gases recognised under the Kyoto Protocol. The numbers in brackets after
each gas represent the IPCC's estimate of the 100-year global warming potential for each gas,
expressed as an equivalent mass of carbon dioxide; for example, a quantity of methane has 21 times
the warming effect of the equivalent amount of carbon dioxide.

Strangely, too, the most important greenhouse gas of all, water vapour, is missing both from this list and from the public debate. This omission is justified by the Intergovernmental Panel on Climate Change (IPCC) on the grounds that the lifetime of water vapour in the atmosphere is only a few days, making it a climate feedback rather than a forcing agent[35]. This argument exploits a dubious technical nicety, and also depends upon the IPCC's estimate of longevity for carbon dioxide in the atmosphere of more than one hundred years. This estimate conflicts with a large number of other, independent estimates of carbon dioxide lifetime, which range between five and ten years[70].

Carbon dioxide is referred to as a 'greenhouse' gas because – together with water vapour, methane, nitrous oxides and ozone – it has the effect of absorbing, and then re-emitting, Earth's spacebound infra-red heat radiation, thereby producing atmospheric warming. The combined effect of these greenhouse gases is generally argued to warm Earth's atmosphere by 34°C, from a chilly, theoretical -19°C in their absence to a pleasant (to us) +15°C in their presence[71].

A typical estimate of the proportions of the Earth's estimated 34°C of warming that is caused by the main greenhouse gases is 78 per cent from water vapour, 20 per cent from carbon dioxide and 2 per cent from other minor gases including methane and nitrous oxide[71]. Higher (88 per cent)[72] and lower (60 per cent)[73] estimates exist for the percentage contributions of water vapour, but all researchers agree that it is clearly the dominant and most important greenhouse agent.

The IPCC argues that about half of the 7 Gigatonne (10^9 tonnes; Gt) of carbon (as carbon dioxide) that humans are estimated to add currently to the atmosphere accrues there each year to become part of the ~780 Gt of total carbon now present. Other calculations described below, and based upon a shorter carbon dioxide residence time and carbon isotope evidence, suggest that only four to five per cent of present atmospheric carbon dioxide is derived from fossil-fuel burning. But even using the IPCC's likely exaggerated estimate, human emissions would still only account for 0.45 per cent (3.5 parts of 780) of the greenhouse warming in a particular year, which is similar to the 0.1°C of possible human warming calculated by Segalstad (1996)[74]. Put another way, 99.55 per cent of the greenhouse effect has nothing to do with human carbon dioxide emissions. These facts notwithstanding, and despite water's combined gas+liquid+solid greenhouse dominance over planetary warming, carbon dioxide remains at the centre of the public debate and so it is to carbon dioxide that we will now turn in more detail.

The CO_2 villain

Public discussion about 'carbon (sic) policy' or 'reducing greenhouse gases' centres on the perceived need to reduce human emissions of carbon dioxide. In turn, that need is a response to the presumption that increased, and still increasing, human emissions of carbon dioxide will cause dangerous climatic warming beyond the natural 34°C envelope.

It is widely asserted that atmospheric carbon dioxide levels have increased steadily from about 290 parts per million (ppm) to 380 ppm since 1900 (32 per cent), and there is little dispute among scientists that human emissions are one of the causes. Nor is there any disagreement that carbon dioxide is a greenhouse gas that exerts a small initial warming effect on the climate system, these points and other related issues being well covered in the thorough reviews by de Freitas (2002) and Soon (2007)[75].

Beyond this, however, it is disputed whether the twentieth century rise in carbon dioxide was really steadily progressive[76]. For example, Slocum (1955)[77] demonstrated an average level of 335 ppm of atmospheric carbon dioxide, with little variation, over the last two centuries, and Ernst Beck (2007)[78] has summarized some 200,000 wet chemical analyses of atmospheric carbon dioxide from the nineteenth and twentieth centuries which show irregular fluctuations that include a 380 ppm peak in atmospheric carbon dioxide levels in the 1940s.

In addition to the uncertainty arising from these historical measurements, there is also vigorous debate as to the magnitude of any warming effect of human carbon dioxide emissions once all likely feedback loops are considered. Important points that relate to this debate include the following.

The global carbon dioxide budget: human contributions in context
The effect of human emissions on global levels of atmospheric carbon dioxide is not well understood because no one, including the IPCC, can satisfactorily account for the observed levels in detail; our best estimates of carbon dioxide sources and sinks have large error bars.

For example, until recently estimates of the carbon dioxide yield of one of the world's best known land volcanoes, Kilauea Volcano (Hawaii), was 2,800 tonnes/CO_2/day. In 2001, Gerlach and co-authors established by measurement a more accurate figure of 8,800 tonnes/day, which is over three times as great[79]. If such uncertainty attends to well-studied subaerial volcanoes, the estimates of carbon dioxide emissions from submarine volcanoes, the majority, are obviously little better than guesses. As another example, in 2006 a newly discovered seabed volcano southeast of Japan was described to be venting copious quantities of liquid carbon dioxide, leading to the dry but entirely accurate comment that 'submarine arc volcanoes may play a larger role in oceanic carbon cycling than previously realized.'[80] One volcanic onland super-eruption is also capable of injecting a huge amount of carbon dioxide into the atmosphere[81].

It is clear, therefore, that the global carbon dioxide budget is poorly constrained. With this caveat, what we think we do know includes the following.

The atmosphere contains about 780 Gt of carbon, of which ~90 Gt is exchanged each year with the oceans and another ~120 Gt with plants. Thus about 25 per cent of the atmospheric carbon is turned over each year, and the observed decrease in radioactive carbon ^{14}C after the cessation of atmospheric nuclear bomb tests in

1963 confirms that the half-life of carbon dioxide in the atmosphere is less than ten years[82]. The residence time, at around five years, is much shorter than assumed by the IPCC[70]. In addition, the ocean has about 39,000 Gt of carbon dissolved in it, some of which is sequestered each year in the formation of seafloor limestone; 70 million Gt of carbon has accrued cumulatively in limestones and other carbonate rocks through geological time, and soils, vegetation and humus contain another 2,300 Gt[83]. Carbon is constantly exchanged between these and other reservoirs, among which the small modern atmospheric reservoir is carbon-starved by comparison with earlier geological history (Fig. 14). The ocean-atmosphere interface is particularly active in carbon exchange, and as the vapour pressure of carbon dioxide above the ocean rises with ocean temperature, the oceans eventually absorb any carbon dioxide above the equilibrium vapour pressure. Man's carbon dioxide contribution is small in the context of the planetary carbon system, but the IPCC argues that, nonetheless, anthropogenic emissions will 'tip' the natural balance of the planet, causing dangerous climate change and acidification of the ocean.

One estimate, by Canadian climatologist Tim Ball, is that the human production of carbon dioxide (7.2 Gt C/year; IPCC, 2007[5]) is more than four times less than the combined error (32 Gt) on the estimated carbon dioxide production from all other sources[84]. A perspective that follows is that even were human emissions to be reduced to zero, the difference would be lost among other uncertainties in the global carbon budget. What is presently missing from the public debate, then – and it is not provided by computer model outputs, either – is an appreciation of both the small scale (in context) of human emissions, and the range of uncertainty in the carbon budget.

In light of these numbers and rapid rates of carbon turnover, the small observed trends in absorption of carbon dioxide attributed to humans should not be of serious concern. In the long run, when equilibrium is achieved, human-caused emissions will have an insignificant effect on the amounts of carbon dioxide in

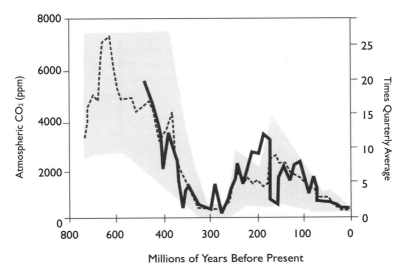

Fig. 14 Reconstructed planetary carbon dioxide levels for the last 600 million years, determined at 10 million year intervals. Note the overall trend for diminishing atmospheric carbon dioxide over time, and the steep drop after the emergence of land plants about 400 million years BP. The planet today is in a state of carbon dioxide starvation compared with levels during most of geological time. Bold black line: average from 372 measurements of palaeo-atmospheric proxies; dotted line and flanking shaded error zones: projections made with the GEOCARB III model by Berner & Kothavala (2001).

the atmosphere and oceans. Though we know little about the transient effect of human emissions, there is little reason to suspect that the effect is dangerous. Indeed, as covered in more detail below, enhanced carbon dioxide levels are clearly beneficial for both agronomy and for plant growth more generally.

Less temperature bang for every carbon dioxide buck
When told that adding carbon dioxide to the atmosphere causes increased temperature, because of the greenhouse effect, most lay persons presume that the relationship is a directly linear one, i.e., that the temperature increases in steady and equal increments as carbon dioxide is added. But greenhouse gases aren't like that; instead, as extra gas is added to the atmosphere, incremental

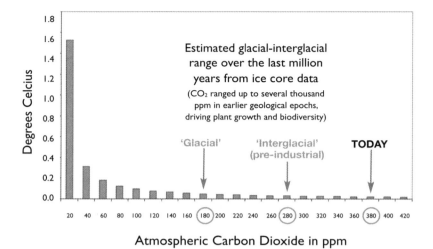

Fig. 15 Model projection of the warming effect of injecting carbon dioxide into Earth's atmosphere, plotted in 20 ppm increments. Note the logarithmic nature of the resulting warming pattern, with most of the warming (2°C) being caused by the first 60 ppm of carbon dioxide. Based on the MODTRAN atmospheric model, University of Chicago.

temperature increases occur in declining logarithmic fashion – as the heading above says in everyday terms. A good analogy is of painting the windows of a room with white paint: the first coat or two of paint block much of the light and cause a marked darkening of the room, but by the time the tenth coat and more are added, little additional darkening is observed. Similarly, the first 60 ppm of carbon dioxide added to the newly forming Earth atmosphere is modelled as causing 2°C of warming, whereas adding another 60 ppm from, say, 300-360 ppm causes less than one-tenth of a degree of additional warming (Fig. 15).

The critical point to grasp is that the negative logarithmic relationship that exists between the addition of carbon dioxide to the atmosphere and consequential radiative heating means that each incremental amount of extra carbon dioxide exerts a *lesser* heating effect (IPCC, 2001, p.635[5]). This relationship provides a perfect example of the law of diminishing returns. The 100 ppm post-industrial increase in carbon dioxide from ~280 to 380 ppm must therefore have already caused most (about 75%) of the anticipated 1.2°C of human warming which supposedly is being caused by a doubling of carbon dioxide from its pre-industrial level[85]. So even if one concedes that about 1°C of warming would occur for a doubling (see next section), all that remains to occur for a completion of doubling now is additional warming of an insignificant few tenths of a degree.

Climate sensitivity to carbon dioxide doubling is low
IPCC models predict much greater increases in temperature for a doubling of carbon dioxide than the 1°C mentioned above, up to 6.4°C (IPCC, 2001) or 4.5°C (IPCC, 2007)[5]. This is because the IPCC's calculations selectively take account of positive feedback effects that further increase temperature (for example, water vapour), whilst at the same time neglecting negative feedback loops that act to decrease temperature (such as the generation of additional low cloud cover, which reflects incoming solar radiation back to space). As Richard Lindzen has said, the idea that the climate system 'is dominated by positive feedbacks is intuitively implausible, and the history of the Earth's climate offers some guidance on this matter[86].'

Supporting this intuition, many research papers by independent scientists, several of which are based on observational evidence, suggest that, contrary to IPCC belief, the net climate feedback of doubled carbon dioxide will be at most a few tenths of a degree of warming (Fig. 16)[87].

Increase in atmospheric carbon dioxide shows a poor correlation with temperature
It is implied in many alarmist writings that a close correlation exists between the twentieth century increase in atmospheric carbon dioxide and global average temperature. This assertion is untrue despite the fact that the twentieth century did witness an overall increase in both temperature (at most by 0.7°C) and carbon dioxide

(from 300 to 380 ppm) (Fig. 10). In detail, however, the two curves are very different, and include the conspicuous mismatch that carbon dioxide recorded its highest rate of increase between 1940 and 1975, at almost precisely the time that global temperature decreased for three decades. Attempts to explain this embarrassing anomaly away by reference to aerosol-induced cooling are simply special pleading.

In contrast to the mismatch between carbon dioxide and temperature, multi-decadal temperature cycles in the Arctic region show a close correspondence to identifiable fluctuations of solar activity (Fig. 17)[88], consistent with the idea that solar changes are the dominant driver of regional if not global climate.

The urban carbon dioxide dome test

The intensity of industrial and vehicular production of carbon dioxide often leads to the development above cities of a concentrated dome of carbon dioxide, especially in winter when the lack of photosynthesizing vegetation leads to a reduced sink for the gas[89]. For example, winter levels of carbon dioxide above Phoenix, Arizona, range up to 500 ppm and commonly exceed 400 ppm. That the winter temperatures for Phoenix show no sign of any warming as a result of this is another specific, and negative, test of the IPCC's hypothesis of dangerous warming caused by carbon dioxide emissions[90].

The disputed evidence from carbon isotopes

Isotopes are varieties of atoms of an element whose nucleii have the same number of protons but different numbers of neutrons. Thus different isotopes exhibit nearly identical chemical behaviour, but because of their slightly differing masses they may become preferentially enriched or depleted when participating in physical processes.

Carbon has three isotopes, ^{12}C, ^{13}C and ^{14}C. Many readers will be aware that because ^{14}C is radioactive, i.e. breaks down over time, its diminishing abundance in fossil materials can be used as

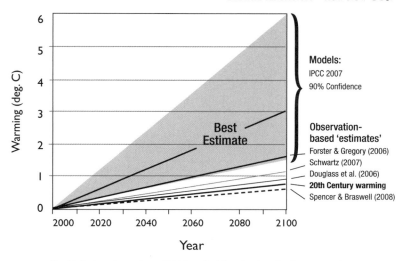

Fig. 16 Projected warming to 2100 for a doubling of carbon dioxide (termed climate sensitivity), as indicated by IPCC models (1.6-6°C; grey field) and observational indicators reported by independent researchers (0.6 to 1.2°C; grey lines).

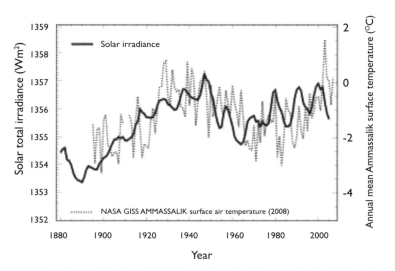

Fig. 17 Comparison between annual total solar irradiance (solid line) and measured surface air temperature at the coastal Greenland location Ammassalik for the period 1881-2007. Note the close correspondence between the two records, consistent with a direct link between temperature and solar irradiance at this locality.

a means of estimating their ages back to about 50,000 years ago (beyond which age, too little of the original ^{14}C remains to allow accurate measurement). Regarding the two stable carbon isotopes, however, many photosynthesizing plants selectively draw in carbon dioxide that is enriched in lighter ^{12}C, resulting in plant tissues that are depleted in ^{13}C by about 1.8 per thousand (per mil) in comparison with the natural ratio in the atmosphere. In contrast, inorganic carbon that is dissolved into the oceans is enriched by about 0.7 per mil compared to the atmosphere. Thus individual carbon reservoirs such as the atmosphere, the ocean and plant biomass each have their own characteristic carbon isotope signature. That signature is expressed as a delta ratio between the two stable isotopes, ^{13}C and ^{12}C, in parts per mil with respect to an internationally recognized reference standard (PDB). In these terms, the atmosphere had a delta value of -7 per mil prior to the industrial revolution, the ocean reservoir (dominated by the bicarbonate ion) has a ratio of zero per mil, and fossil fuel and terrestrial biogenic material possesses a delta value of -26 per mil (Fig. 18).

The burning of fossil fuels or forest vegetation releases light carbon into the atmosphere, thereby decreasing the ^{13}C/^{12}C ratio by measurable amounts. Study of samples of ancient atmospheric carbon dioxide from ice cores since 1800 show an initial decline in ^{13}C/^{12}C ratio in 1850. Modern flask sampling of the atmosphere shows a steeper rate of decline after 1960[91], with a change from -7.489 per mil in 1978 to -7.807 in 1988[92]. These changes to some degree correspond with independent estimates of the rates of human production of carbon dioxide due to industry, transport and land clearing. Related evidence includes parallel declines in the ^{14}C/^{12}C ratio of atmospheric carbon dioxide (the 'Suess effect', which occurs because fossil fuels contain no radioactive ^{14}C) and in the oxygen concentration in the atmosphere (a signature of the oxidation of carbon by burning of fossil fuels or the decay of organic matter)[93]. These geochemical arguments are viewed by the IPCC and its supporters as 'killing proofs' in support of a human origin for

Atmospheric Carbon Dioxide in ppm

Onland (organic, 2,300 Gt)
Trees (C3); 1,090 Gt, $\delta^{13}C$ = -25‰
Grasses (C4); 60 Gt, $\delta^{13}C$ = -13‰
Soil, peat; 1,150 Gt, $\delta^{13}C$ = -26‰

Onland (human, 7.2 Gt)
Fossil fuels, cement; 7.2 Gt, $\delta^{13}C$ = -26‰

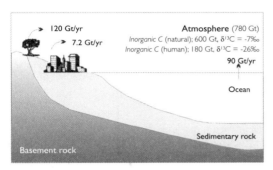

Surface ocean (1000 Gt)
Inorganic C; 975 Gt, $\delta^{13}C$ = +1‰
Organic C; 25 Gt, $\delta^{13}C$ = -25‰

Deep ocean (38,000 Gt)
Inorganic C; 37,200 Gt, $\delta^{13}C$ = 0‰
Organic C; 800 Gt, $\delta^{13}C$ = -22‰

Sequestered in sediments (inorganic)
Methane; (CH₄), $\delta^{13}C$ = -50 to -70‰
Oil; $\delta^{13}C$ = -21 to -34‰
Coal (C); $\delta^{13}C$ = -24 to -34‰
Carbonate (Ca, MgCO₃); 70 million Gt, $\delta^{13}C$ = +2‰

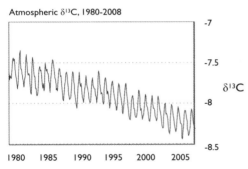

Fig. 18 Major reservoirs of carbon on Earth have characteristic amounts of carbon (in Gt) and characteristic carbon isotope $\delta^{13}C$ values. Analysis of the changing isotope ratio over time (lower diagram) can therefore be used to discriminate between sources for the carbon in atmospheric carbon dioxide.

most of the estimated post-industrial ~100 ppm rise in atmospheric carbon dioxide (IPCC, 2001, Section 3.3.3 5)[94].

IPCC scientists also use isotope analysis to estimate the proportion of human emissions that remain within the atmosphere each year. With a natural flux each year of 210 Gt/C/yr and a 7.2

Gt/yr human addition, the fossil fuel isotope label should be 3.4 per cent (7/210) of the total shift observed, whereas in fact it is only about half of that. From this the IPCC concludes that about half the human input each year is absorbed by an unknown or 'missing' carbon sink, leaving the other half to accumulate in the atmosphere. This reasoning is, however, flawed, because it assumes the existence of constant equilibrium, and ignores the possibility that the atmospheric signature might stem from dilution by incremental carbon dioxide additions from other natural sources.

Not all scientists agree that the isotopic arguments are conclusive. First, because soil and forest carbon, like fossil fuel carbon, are strongly depleted in ^{13}C, and therefore feasible sources that may be partly driving the $^{13}C/^{12}C$ decline. And, second, because the IPCC's calculations involve estimates of the residence time for carbon dioxide in the atmosphere, which is often quoted to be 50-200 years (IPCC, 1990[5]) or longer (Archer et al., 2009[95]), and which Princeton Professor Harvey Lamb has shown actually averages 400 years for all published IPCC models up to the third assessment report[96]. Abundant published research shows that such a figure is inflated by two orders of magnitude from the 5-15 year estimates of independent scientists[70]. Allowing for the real, shorter lifetime in his calculations, and using different techniques, Norwegian geochemist Tom Segalstad[97] has independently estimated that only about four per cent of the current atmospheric carbon dioxide is derived from human and fossil fuel sources. US meteorologist Roy Spencer, using similar reasoning to construct a simple model[98], has also concluded 'that most of the upward trend in carbon dioxide since ... monitoring began at Mauna Loa 50 years ago could indeed be explained as a result of the warming, rather than the other way round.' These studies imply that the natural processes of marine outgassing of carbon dioxide with temperature rise and juvenile outgassing from volcanic sources must be much more important, and the burning of fossil fuels less important, in contributing to the current rise in carbon dioxide than is argued by the IPCC.

Since when does an effect precede its alleged cause?

One of our great research resources for understanding recent climate change is the data available from ice cores drilled in Antarctica (a record back to almost one million years; Fig. 3) and Greenland (back to about 0.13 million years; Fig. 4)[99]. Uniquely, ice cores trap bubbles of ancient atmospheric gas, which allows the measurement of carbon dioxide and methane levels long ago. During the natural Milankovitch-scale climate cycles that the cores reveal, the changes in temperature always precede the parallel changes in carbon dioxide[22], which is quite the opposite of the misleading impression given by, for example, Al Gore's film, *An Inconvenient Truth*. In addition, and as would be expected if ocean outgassing is a primary source for enhanced interglacial carbon dioxide levels, similar research on Pacific marine cores shows that 'deep-sea temperatures warmed by ~2°C between 19 and 17 thousand years before present, leading the rise in atmospheric CO_2 and tropical surface ocean warming by ~1000 years[100].' Thus the evidence is clear that atmospheric carbon dioxide level is not the primary driver of global temperature change at the scale of glacial/interglacial cycles, but rather a consequence of the temperature change.

Other research has traced the seasonal changes in atmospheric carbon dioxide that occur as deciduous trees in the northern hemisphere wax and wane with the seasons, causing atmospheric carbon dioxide to rise during the autumn and winter (as plants shed their leaves, and cease photosynthesizing) and fall during the spring and summer (as regrowth occurs). This research again finds that change in temperature precedes the parallel change in carbon dioxide, this time at a short, annual time-scale[101]. As the heading says, it is a strange cause that thus post-dates its supposed effect.

Carbon dioxide is the staff of life

Carbon dioxide has previously reached concentrations similar to today's industrially enhanced levels a few thousand years ago, in the early Holocene[102]. Prior to that, in earlier geological epochs, atmospheric carbon dioxide attained levels of 1,000 ppm and much more without known untoward environmental effects (Fig. 14)[103].

Increasing atmospheric carbon dioxide both enhances plant growth and aids the efficiency of water use. Therefore, enhanced carbon dioxide is of net benefit for both biodiversity and food production[104]. For example, a recent study by Smithsonian Institute ecologist Sean McMahon and colleagues showed that historical changes in temperature and carbon dioxide have resulted in 'recent biomass accumulation [that] greatly exceeded the expected growth caused by natural recovery' from forest disturbance[105]. And, regarding agronomy, when grown under the 560 ppm conditions of double the presumed pre-industrial carbon dioxide level, common food crops deliver the following increases in yield: wheat, 60 per cent; legumes, 62 per cent; other cereals, 70 per cent; fruits, 33 per cent; tuber crops, 67 per cent; and vegetables, 51 per cent, for an overall average of 57 per cent. Far from being 'dangerously' high, at 390 ppm today's level of atmospheric carbon dioxide is well below both the nutrient saturation level and the optimal level for growth for many plants[106].

During the second half of the twentieth century, human population increased 2.2 times, at the same time that world food production increased by 2.7 times. Accordingly, many more people were fed, and many others fed better, by what has been termed the 'green revolution'. Though better agronomy and plant breeding underpinned most of the increase in food production, the productivity boost was assisted also by a 10-15 per cent fertilisation effect of increasing carbon dioxide in the atmosphere[107].

Conclusion
The factual information provided in this chapter indicates that only minor warming will result from further increases in atmospheric carbon dioxide above the presumed pre-industrial level of 280 ppm. It follows that cutting carbon dioxide emissions, be it in single countries or worldwide, is unlikely to cause any measurable difference in future climate.

Neither does any case exist for the assumption that higher levels of carbon dioxide are, of themselves, harmful. First, because any mild warming caused by enhanced carbon dioxide is likely to be of net climatic benefit; and, second, because higher atmospheric carbon dioxide both enhances plant growth and aids efficiency of water use. In reality, enhanced atmospheric carbon dioxide is a net benefit for biodiversity, food production and greening of the planet.

I conclude that the widespread claim that higher levels of atmospheric carbon dioxide are, of themselves, causing dangerous warming, or are otherwise environmentally harmful, is simply untrue. To term carbon dioxide a pollutant, as did a 5:4 majority of the US Supreme Court in the 2007 case of Massachusetts versus the Environmental Protection Agency (EPA)[108] – thereby delivering public regulation of emissions into the hands of the EPA, where (legally) it still resides – is an abuse of language, an abuse of logic and an abuse of science.

Charles Krauthammer[109] has well foreseen where this approach will lead should it remain unchecked:

> On the day Copenhagen opened, the US Environmental Protection Agency claimed jurisdiction over the regulation of carbon emissions by declaring them an 'endangerment' to human health. Since we operate an overwhelmingly carbon-based economy, the EPA will be regulating practically everything. No institution that emits more than 250 tons of carbon dioxide a year will fall outside EPA control.
>
> This means over a million building complexes, hospitals, plants, schools, businesses and similar enterprises. (The EPA proposes regulating emissions only above 25,000 tons, but it has no such authority.) Not since the creation of the Internal Revenue Service has a federal agency been given more intrusive power over every aspect of economic life.

This naked assertion of vast executive power in the name of the environment is the perfect fulfilment of the prediction of Czech president (and economist) Vaclav Klaus that environmentalism is becoming the new socialism, i.e. the totemic ideal in the name of which government seizes the commanding heights of the economy and society.

4   Ocean matters: hobgoblins of alarm

> The whole aim of practical politics is to keep the populace alarmed (and hence clamorous to be led to safety) by menacing it with an endless series of hobgoblins, all of them imaginary.
>
> (H. L. Mencken)

In this chapter we will deal with two ocean topics that have been endlessly hyped to the public as of global warming concern, namely sea-level rise and the 'acidification' of the ocean. These two matters need to be considered against a background understanding of the vastness in size of the modern ocean, of its thermal inertia, and of the degree to which ocean heat and dissolved gases, as they exchange with the atmosphere, are major determinants of climate.

The world ocean

The heat capacity of the ocean

The ocean covers more than 70 per cent of the Earth's surface, and over much of its area it is 3-5 km deep. Comprising water, which is one thousand times denser than air, the ocean has far more mass than the atmosphere (1.5×10^{18} tonnes compared to 5×10^{15} tonnes) – notwithstanding that the atmosphere covers the entire planet and is 50 km high to the top of the stratosphere. The result of this is that the ocean has a much greater heat capacity than the atmosphere, specifically 3,300 times more. Put another way, all the heat energy contained in the atmosphere is matched by the heat content of only the upper 3.2 metres (m) of the worldwide ocean[110].

Another consequence is that water requires much more energy to heat it up than does air. On a volume/volume basis, the ratio of heat capacities is, of course, 3,300 to 1. One practical result of this is that it is almost impossible for the atmosphere to exert a significant heating effect on the ocean, as is often asserted to by promoters of global warming alarm. For to heat one litre of water by 1°C will take 3,300 litres of air that was 2° hotter, or one litre of air that was 3,300°C hotter, neither of which is a common scenario within our every day weather system. Instead, it is the ocean that controls the warmth of the lower atmosphere, in three main ways: namely, through direct contact, by infrared radiation from the ocean surface and by the removal of latent heat by evaporation.

Leads and lags

Earth's climate is all about the reception and transport around the planet of heat energy received primarily from the sun. Despite the overwhelming concentration on atmospheric weather processes in the public climate debate, it is really the ocean that calls the biggest shots in determining how the heat we receive from the sun is absorbed, recirculated and re-emitted over climatic time-scales. The time constant for the atmosphere, during which a molecule of carbon dioxide may be circulated worldwide, is about one year; the time constant for the ocean, during which a message in a bottle released in Antarctica might be circulated by worldwide ocean currents to pitch up on Greenland's icy shores, is one thousand years and longer.

The practical effect of this is that major time lags are built into the climate system, such that a warming or cooling event that occurs today (say the Great Pacific Climate Shift in 1976-77, which corresponded to a worldwide step-increase in temperature of about 0.2°C) may be reflecting a change in heat energy that was stored in the ocean many hundred years ago, and which for one reason or another is only being released now – by cold water upwelling from the depths of the ocean, for example. To give another example, some scientists suggest that part of the rise in atmospheric carbon dioxide that has occurred in the twentieth century may represent ocean outgassing caused by the Mediaeval Warm Period, after the same time lag of around 1,000 years that is observed in the ice core record.

Changes in the heat content of the ocean are important from a human perspective for three reasons. First, because the ocean is coupled to the atmosphere, there is an effect on atmospheric temperature and therefore on the equator-to-pole thermal gradient that drives climate processes. Second, because of the decreasing solubility of carbon dioxide with increasing temperature, oceanic ingassing or outgassing caused by changing ocean temperature modulates the level of greenhouse gas in the atmosphere, affecting climate again. Finally, as the ocean heats or cools it expands and contracts commensurately, and thereby has an effect on sea-level called the *steric* effect.

Sea-level change

It is widely assumed that sea-level is rising uniformly all around the world, and probably at dangerous rates to boot. A second public belief is that the Intergovernmental Panel on Climate Change (IPCC) provides predictions about future sea-level rise that are useful for coastal management.

Regarding the first assumption, historical measurements made with tide gauges, and geological observations that allow historic sea-level to be reconstructed, have established firmly that the rate of *local* sea-level change varies widely around the world, rising in many places and falling in others. And regarding the second, the IPCC has never claimed to provide validated *predictions* of future sea-level rise, but rather furnishes governments with *projections* from speculative computer models and socio-economic scenarios that are known to be flawed[111]; in other words the IPCC deals with virtual reality futures only (see Chapter 9).

Both of the opening assumptions are therefore wrong, as summarized well in this quotation from the Nongovernmental International Panel on Climate Change (NIPCC) (2008, p. 51)[112]:

> Sea-level rise is one of the most feared impacts of any future global warming, but public discussion of the problem is beset by poor data and extremely misleading analysis.... Most discussion, including that of the IPCC, is formulated in terms of global average [eustatic] sea-level. Even assuming that this statistic can be estimated accurately, it has little practical policy value. Local relative sea-level change is all that counts for purposes of coastal planning, and this is highly variable worldwide, depending upon the differing rates at which particular coasts are undergoing tectonic uplift or subsidence. There is no meaningful global average for local relative sea-level.

As is often the case for issues where science and politics mix, the reason for such widespread public misunderstanding of some elementary science facts is sloppy (and perhaps deliberately sloppy) terminology linked to inadequate science understanding.

Global or local sea-level change?
A clear understanding of the risk associated with sea-level change requires the use of accurate terminology. In particular, a sharp distinction must be drawn between changes in *eustatic* sea-level, a notional world-wide average, and changes in *local relative sea-level* (LRSL), which correspond to changes in actual sea-levels at real and particular coastal locations.

In general, statements made by the IPCC and by government planning and management authorities which use the unqualified and ambiguous term 'sea-level' are referring to global average sea-level. Our current estimate of eustatic sea-level for the last 20,000 years, based on studies and samples from around the world, is shown in Fig. 19. On its own – without taking into account the factors of local uplift or subsidence of the land substrate and local sediment supply – this knowledge of global sea-level behaviour does not enable the determination of future shoreline positions in particular places, and, as the NIPCC quotation above points out, it is

Post-Glacial Sea Level Rise

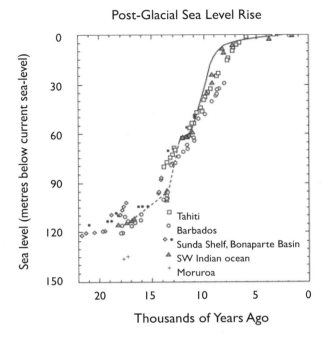

Fig. 19 Global (eustatic) sea-level since the peak of the last glaciation, 22,000 yr before present, as delineated by geological sampling and dating. Note the very rapid melting at rates up to 2m/century between about 15,000 and 10,000 years ago, after which the rate of rise lessens to 1-2 mm/yr in the Holocene, a gentle rate of rise that has continued over historic times.

therefore of little value for coastal management and environmental protection purposes.

Architects who work in Calgary, Jakarta and Phoenix design heating and cooling systems for their clients' houses based on the local climate, not on some mythical global average climate, or at least they do if they want to stay in business. And so it is for coastal management, which engineers have traditionally undertaken by using real knowledge of the real-world, local sea-level and geological conditions.

The alternative, post-modern approach to coastal engineering favoured by the IPCC seems unlikely to have much of a future, for the reality is that LRSL change occurs at greatly different rates and directions at different coastal locations around the world. Locations that were formerly beneath the great northern hemisphere ice caps 20,000 years ago, having been depressed under the weight of the ice, started to rise again as the ice melted, and that rise continues today, for example in Scandinavia at rates up to 9 mm/year. Accordingly, *local relative sea-level* in such areas is now falling through time despite a concurrent long-term rise in *eustatic (global) sea-level* over recent centuries of about 1.7 mm/year. Conversely, at far field locations distant from polar ice caps, such as Australia, no glacial rebound is occurring and rates of sediment supply are low, which results in local sea-level change in most places being controlled by the global average rise; therefore, at many but not all locations around the Australian coast, sea-level has risen over the last century at rates between 1 and 2 mm/year[113].

But why is it that the global sea-level rise is not mirrored exactly at every location around the world's continental coastlines? The answer is that the geological substrates differ from place to place and are rarely absolutely stable; rather, the substrate is sinking in some places (for example on delta coasts, such as around the Gulf of Mexico or in Bangladesh) and rising in others (for example in many, but again not all, places in earthquake-prone countries like Japan or New Zealand). As a result, around the world we see a complex pastiche of different rates of local sea-level rise or fall, with the position of the shoreline at any one time and place being dependent on the interaction of three things – the rate of substrate movement up or down, the rate of sediment supply from rivers versus marine transport or erosion, and the rate of change in eustatic sea-level.

Sea-level estimates from tide gauge measurements
Over historic time-scales up to three hundred years, sea-level can be measured directly using tide gauge records (since 1700 in Amsterdam; Fig. 20). Corrections are applied for substrate sinking or uplift, and for oceanographic or meteorologic distortions of the

Sea Level has risen by 10–20 cm/century

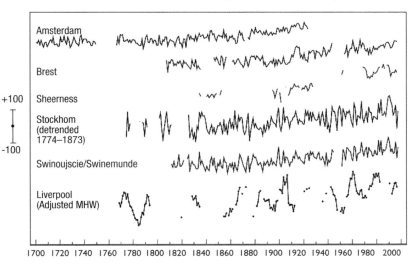

Fig. 20 Selected long-term tide gauge records of sea-level from Northern Europe, corrected for local tectonic effects. The scale bar indicates 100 mm of rise or fall. The average residual rate of sea-level rise during the twentieth century is ~1.8 mm/yr, with no observable recent acceleration.

underlying global signal (for example, any increased air pressure, which serves to depress sea-level, or any increase in speed of the major ocean current gyres, which acts to increase sea-level around their periphery), to derive an estimated global signal. After making such corrections, the best long-term tide gauge records measure a twentieth century average sea-level rise of 1.74 mm/year[114], and recent GPS-corrected measurements indicate a rate of 1.61 mm /year[115]. IPCC (2007)[5] estimates a similar figure of 1.8 mm/year, and this gentle rise of 1.6-1.8 mm/year simply represents the continuation of the last stages of the sea-level rise caused by the melting of the ice sheets since about 17,000 years ago (Fig. 19). IPCC estimates that their 1.8 mm/year rise is partitioned into 0.4 mm/year caused by thermal expansion of the ocean, 0.7 mm/year contributed by ice melt and a residual, unaccounted 0.7 mm/year (which may relate to dynamic oceanographic and meteorological factors).

Short-term sea-level estimates from satellite radar altimetry

Since 1992, TOPEX and POSEIDON mission satellites equipped with radar altimeters have been able to measure sea-level with great accuracy. Global sea-level maps can be constructed every ten days with a claimed uncertainty of only 3-4 mm. In addition, the satellite data from specific locations around the world can be used to construct accurate records of local sea-level change.

Prima facie, the modern satellite altimetry measurements indicate a rate of global mean sea-level rise of 3.2 mm/year between 1992 and 2009[116], which is roughly twice the rate inferred from the tide gauge records. The reason for the difference is not fully understood, but probably at least partly relates to the complex corrections that have to be made to the satellite data, which cannot yet be viewed as securely accurate. First, because of the limited accuracy of the measurements (nominally about +/- 100 mm, which can be improved to about +/- 20 mm by averaging ten day-separated repeat measurements at particular locations); second because of corrections that are needed for orbital drift and the stitching of records from different satellites, and third because of the short length of record of only about 15 years, whereas a minimum 40-50-year length of record is required to delineate a meaningful sea-level trend.

The calibration and correction of the radar altimetric sea-level records is therefore still under discussion. For example, US oceanographer Carl Wunsch and colleagues[117] present general circulation modelling (GCM) that, taking account of hydrography, sea surface temperature and other regionally variable factors, yields a global mean increase in sea-level of ~1.6 mm/yr for the period 1993-2004, which corresponds to the figure estimated from tide gauge records, and is only about 60 per cent of the altimetric estimate. These authors conclude that: 'At best, the determination and attribution of global-mean sea-level change lies at the very edge of knowledge and technology.'

The perils of fitting straight lines through cyclic data

Sea-level changes on a decadal or multi-decadal scale are driven by changes in the heat energy or dynamics of the ocean system. They include the effects of evaporation (which causes a semi-permanent low over the tropical Indian Ocean), the spinning up or slowing down of major current gyres, and the effects of established climatic oscillations such as the Atlantic Multi-decadal Oscillation (AMO), El Nino-Southern Oscillation (ENSO), the Pacific Decadal Oscillation (PDO) and the Southern Annular Mode (SAM). Sea-level change forced by such mechanisms generally has low magnitudes of centimetres to a metre or two, but it can operate at rates as high as 5-10 mm/year. Another important cause of shorter term changes in sea-level is that produced by heating or cooling of the ocean. After a phase of warming, expansion and steric sea-level rise during the twentieth century, ocean cooling has led to steric sea-level fall since around 2004[118].

Because of these dynamic effects, and like global temperature, sea-level change is periodic on a multi-decadal scale, so fitting straight lines through the data is an unreliable technique with which to infer trends for use in environmental management. Calculations of trends in sea-level measurements are highly sensitive to the start and end points of the dataset being considered (Chapter 2, Fig. 13; *How long is a piece of climate string?*), and most such estimates ignore the short-term and multi-decadal changes in sea-level that are known to be associated with meteorological and oceanographic oscillations. For example, the 16 year-long 1993-2009 satellite record can indeed have a trend line fitted through it, but this period is not nearly long enough for the measured trend to be viewed as climatically significant, and that the data exhibit a plateau since about 2004 is consistent with decadal rhythmicity and perhaps an imminent fall in sea-level.

Is global sea-level rise accelerating?

The appropriate question therefore is not 'is global sea-level rising?' for the geological and tide gauge record indicates that over the long term it is, and, other things being equal, will continue to do so. Rather, to provide evidence for increased rates of rise due to human influence necessitates that the question be 'is global sea-level rise accelerating?' And the answer is no.

For example, the IPCC (2001) wrote[5] 'no significant acceleration in the rate of sea-level rise during the twentieth century has been detected'. In 2007 they said[5] that 'global average sea-level rose at an average rate of 1.8 [1.3-2.3] mm per year over 1961 to 2003. The rate was faster over 1993-2003: about 3.1 [2.4-3.8] mm per year. Whether the faster rate for 1993 to 2003 reflects decadal variability or an increase in the longer-term trend is unclear.'

In a recent paper, UK oceanographer Simon Holgate (2008) analysed nine long sea-level records for the period 1904-2003[114]. Holgate's study established that global sea-level rise was 2.03 mm/year between 1904-1953, compared with 1.45 mm/year rise between 1954-2003, which represents a decrease in rate of rise across the century. Similarly, for New Zealand, Hannah (2004) established a long-term rise of 1.6 mm/year for the period 1989-2000, and noted that 'there continues to be no evidence of any acceleration in relative sea-levels over the record period[119].' Many other recent studies have confirmed a lack of observed acceleration in twentieth century sea-level records[120], and most consistently report long-term averages for eustatic sea-level rise that are up to 30 per cent lower than the IPCC's 2007 estimate of 1.8 mm/year, down to 1.2 mm/yr.

Among recent writers, Australians John Church & Neil White (2006)[121] are the only authors who claim to have established that sea-level rise accelerated during the late twentieth century. Their study is based on the same tide gauge dataset used by other authors who have drawn different conclusions, and also conflicts with earlier analyses of the same dataset by the same authors (Church *et al.*, 2004[122]; IPCC, 2007). In any case, any enhanced rate of rise that might have occurred since the early 1990s is just as likely to have been caused by oceanographic as by human forcings.

No cause for alarm

Climate alarmists are masters at the art of exploiting people's fears and guilt. They are aided in this by the advice of expensive public relations consultants, by their attendance at specialized training courses, and by the availability of tactical recipe books – one of the most widely used[123], incredibly, being funded by the US National Science Foundation. Two iconic sea-level scares that have been created using these techniques concern Bangladesh and Pacific coral islands.

Because it is located upon the delta of the Ganges River, much of Bangladesh is low lying. A sea-level rise of only one metre would flood almost 20 per cent of Bangladesh's land area, where about twenty million people live today. Also, because recently deposited muds and sands are water saturated they compact easily, so compaction subsidence and consequent local relative sea-level rise are common in deltaic areas. Lots of people, a large area of vulnerable land and assured sea-level rise – apparently the perfect trio of factors around which to create sea-level alarm. Unfortunately, and dishonestly, the fourth determining factor is always left out of the public discussion, which is that the Ganges delta is currently expanding as sediment supplied by the river accretes along the coast and creates new land. Dhaka geomorphologist Maminul Haque Sarker has shown from analysis of historic maps and satellite images that the delta has been growing in area for centuries, and is currently 'adding nearly twenty square kilometres a year in the coastal areas[124].' What is important, as always, is the balance between subsidence, sediment supply and local sea-level change, and that balance works in Bangladesh's favour. For sure, like Holland (~26 per cent of which lies below sea-level), Bangladesh will inevitably have to expend money on coastal defences for some populated areas on which the sea threatens to encroach, but colonisation onto new land is an equally feasible long-term option, and, anyway, their problem is caused by the dynamics of deltaic sedimentation processes rather than by atmospheric carbon dioxide.

97

The second sea-level pin-up scare is that, as Al Gore asserted in his film, Pacific islander 'climate refugees' – another brilliant linguistic neologism, which hits Western fear and guilt buttons with equal force – are already being forced to evacuate their islands because of rising sea-level. Never mind that a London High Court judgement has found the refugee claim to be untrue (Chapter 6), or that eustatic sea-level exhibits no historically unusual rising trend for most tropical Pacific Islands; once launched into the public domain, a legend such as this one swiftly becomes entrenched courtesy of robotic repetition by green environmental groups and promulgation by negligent media flag-wavers.

In 1993, the Australian government set up a new series of accurate tide gauges to measure sea-level change on 14 selected tropical Pacific islands[125]. Embarrassingly, given the government's expectation that this network would provide information supporting sea-level alarm, inspection of the results reveals an irregularly changing sea-level that is closely correlated with, and driven by, periodic El Nino-La Nina cycling, within which the 1998 super-El Nino event is prominent (e.g. Tuvalu; Fig. 21). This pattern results from most of the measuring sites being located within a warm body of equatorial water called the Western Pacific Warm Pool, within which sea-level is lowered during an El Nino episode by water flowing eastward towards South America, and raised when the water is pushed back westward again during a La Nina. Because the Warm Pool is already at a limiting sea surface temperature of around 30°C, any additional 'global warming' that might be loaded on top of the regional El Nino-Southern Oscillation (ENSO) signal will have little sea-level effect anyway, for the extra energy will be diverted into evaporation rather than causing ocean warming[126].

At 16 years duration, even the longest of the Australian South Pacific Sea-level Project records is of inadequate length to determine accurate long-term trends. Nonetheless, using undisclosed data corrections, rising trends of +3.5-20.6 mm/year are claimed, with Tuvalu allocated a most unlikely rise of 5.3

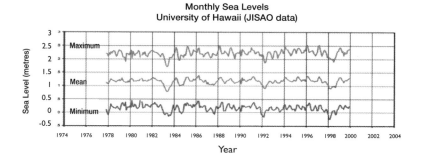

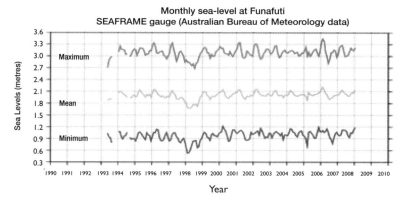

Fig. 21 Two independent sea-level curves for Tuvalu, Pacific Ocean for 1978-2000 (above; after the University of Hawaii, JISAO) and 1994-2008 (below, after the Australian Bureau of Meteorology). Note the irregular fluctuations about a mean position, in sympathy with El Nino-La Nina cycling, but little if any long-term trend. For more detailed explanation, see text.

mm/year. An independent analysis of sea-level change on Tuvalu has been accomplished by Willis Eschenbach[127]. Consulting all available data sources, he compiled four separate estimates of change – namely +0.8, +0.3, +0.07 and -0.8 mm/year – and concluded that:

> The most we can say is that the true long-term rate of MSL [mean sea-level] rise in Tuvalu is very likely to be between

-1 and +0.5 mm/yr, with a best estimate (from the asymptotic analysis of the longest actual record) of +0.07 mm/year. This is well below the lower end of the IPCC estimate of +1 to +2 mm/yr of MSL rise worldwide, and is far too small a rise to be causing the changes which have been reported in Tuvalu.

Nonetheless, it remains the case that, despite the lack of measured evidence, public alarm about Pacific Island sea-level rise is continually maintained by anecdotal accounts (and especially television footage) of local marine flooding from places like Funafuti atoll (Tuvalu), which regularly occurs during storm or king tide events. Well, of course; because Funafuti is an atoll, which means that it is not so much a 'thing' as it is a process. The origin of atolls was famously first explained by Charles Darwin, who noted that they were reef sandbank complexes situated on top of a sinking, extinct volcano. Consequently, atolls are made up of detrital coral sand that has been eroded from living reef communities and washed into shallow banks. Some of these sand banks develop a small fresh-water lens, derived from rainfall, and become temporarily stabilised by vegetation, at which point they can sustain human habitation[128]. Seldom more than a metre or two above sea-level, all atolls and related sand-cay islands are at the continuing mercy of the same wind, waves, tides and weather events that built them. They are dynamic features of the seascape, and over timescales of decades to centuries they erode here, grow there, and sometimes disappear beneath the waves forever. They are obviously not good places in which to develop major human population centres.

The dynamic nature of an atoll is exacerbated, and its integrity even more jeopardized, when it is subjected to the environmental pressures created by a growing human population. Sand mining, construction project loading, and rapid groundwater withdrawal all cause local lowering of the ground surface, and thereby encourage marine incursion quite irrespective of any sea-level change. It is these processes in combination with episodic natural hazards like tides and storms, and not global sea-level change, that provides the alarming television footage that we are now all so used to watching.

No scientific basis for global-warming related sea-level alarm

These elementary facts about sea-level change have been taught to university first year science classes around the world for generations, yet they still seem to elude today's politicians and their advisers. For, also around the world, governments, councils and planning authorities are mostly ignoring the reality of local sea-level change altogether, and instead using the IPCC's speculative projections for future global sea-level as the basis for coastal planning regulations that effectively confiscate property rights and diminish property values[129].

There is no scientific basis whatever for the oft-repeated suggestion that 'global warming' will melt so much ice that sea-levels will imminently rise by Gore's imagined 20ft. The IPCC has successively reduced its best estimate of the sea-level rise to the year 2100 from 0.88 m to 0.59 m to 0.36 m, and nearly all of this decreasing increase arises from the projected thermosteric expansion of the oceans which are, however, currently cooling.

How should we manage sea-level change?

That current and future sea-level change is linked in any measurable way to the influences of mankind on the planetary climate has not been demonstrated, and much of the public discussion on this issue is more akin to science fiction than to sensible science.

That sea-level changes naturally is, on the other hand, a commonplace. Nations like the Dutch have known for centuries that sea-level change can have an adverse impact on humanity, and appropriate adaptive responses have of course been made. There is therefore no question that it is the duty of governments to plan for likely future sea-level changes, as they do for other major natural hazards that affect their nations such as earthquakes or volcanic eruptions. But they need to do this based upon the accurate measurement of sea-level change from their own coastal regions, and the IPCC's speculative computer-based prognostications are not particularly helpful towards that end.

Natural alkalinity variation in the oceans

Since the unusually warm El Nino year of 1998, global average temperature has declined for ten years despite increases in atmospheric carbon dioxide concentration (Chapter 2). Embarrassed by this invalidation of the greenhouse hypothesis, and casting around for another fossil fuel hobgoblin with which to threaten the innocent public, zealous environmentalists have arrived at dissolution of carbon dioxide in the world ocean as a new scare to replace the now discredited scare of dangerous human-caused global warming.

Nothing could make the propaganda intent of this environmental bogeyman clearer than choice of the phrase 'acidification of the ocean' to describe a theoretical change in ocean chemistry that, at its most alarming, will result in a minor decrease in alkalinity of the surface ocean – a change, furthermore, that falls well within the daily and long-term variations that already occur naturally in the ocean.

Acidity, alkalinity and pH

An ion is an atom or molecule that has lost or gained an electron, thus giving it a minute electrical charge that makes it chemically reactive. Common industrial and domestic chemical cleaning compounds may be acidic (with high numbers of positive hydrogen ions, H^+) or alkaline (with high numbers of negative hydroxide ions, OH^-), depending upon the task in hand.

Scientists use a special scale, called the pH scale (for *pondus Hydrogenium* – literally, the weight of hydrogen), to measure the wide 10^{14} range of acid-alkaline ionic conditions that occur in natural and laboratory solutions. The pH scale ranges from 0 (very acidic) through 7 (a neutral solution, e.g. pure water) to 14 (very alkaline). This scale is logarithmic in nature, meaning that the number of hydrogen ions is multiplied by ten for each incremental unit decrease between 14 and 0. Thus a solution with a pH of 1 has 10 times the hydrogen ion concentration of a solution with pH 2, and one hundred times that of a solution with pH 3; and so on.

Solution pH is important for all physico-chemical and also many biological processes. For example, the pH of healthy human blood is slightly alkaline at a pH of 7.3-7.5, and death occurs should blood pH fall below 7.0 or rise above 7.8. Many readers will have determined the pH of solutions during their school education using Litmus paper tests, the colour of the wetted paper being a fairly accurate measure of pH.

Alkalinity of the oceans

The oceans have been alkaline since the late Precambrian, about 750 million years (My) ago, when the amount of carbon dioxide in the atmosphere was up to twenty times higher than now[130]. Today, we live on a carbon dioxide starved planet as judged against geological history (Fig. 14). Since the Precambrian, carbon dioxide has been progressively removed from the atmosphere via inorganic and organic carbonate sediment deposition, mostly in marine environments. Some of these deposits became sequestered later as onland carbonate-rich rocks such as limestone ($CaCO_3$) and dolostone ($CaMgCO_3$) by tectonic processes of accretion of ocean-floor to the continental margins.

These geochemical, geological and biological processes continue today, in an ocean that is now heavily buffered by water-rock and water-sediment reactions[131]. In effect, it is difficult to permanently change the ocean pH by adding acids (including dissolved carbon dioxide, which is normally present as the mildly acidic H_2CO_3) or bases because any increase or decrease in the number of hydrogen ions is first compensated for by reactions with other minerals present, for example clay minerals. At least as long ago as the middle Eocene, ~45 million years ago, the pH range in shallow ocean water is estimated from boron isotope measurements in fossil material to have been similar to today, at 8.33-7.91[132]. This estimate is supported by Holland's (1984) observation[133] that if ancient pH had changed substantially then there would be a considerable effect on the mineralogy of the sedimentary record, which has not in fact occurred. On a cautionary note, however, the use of boron isotopes to determine ancient pH is under challenge[134], because the method is based upon the assumption that the boron isotope inventory of the ocean is a closed system, whereas in actuality rivers constantly supply large amounts of new boron to the ocean; thus measurements of boron isotopes can reflect changes in the boron isotope budget as well as the palaeo-pH of the ocean.

Saturation level

Ocean bio- and geochemical processes that involve carbon dioxide are controlled by the saturation level, which is the maximum amount of gas that can be dissolved in a given volume of sea water. The saturation level varies with temperature, pressure and the concentration of other dissolved materials; pH is consequential and of lesser importance.

Carbon dioxide saturation level decreases with increasing temperature and increases with increasing pressure. Saturation at the ocean surface is therefore controlled mainly by temperature, and at ocean depths near and below 3.5-4.5 km (below a level termed the *lysocline*, below which the rate of calcium carbonate dissolution increases markedly) by pressure. Generally, and as a result of the biological processes of photosynthesis, respiration and decomposition, the oceans at large are undersaturated. So despite the strong buffering, local, regional and depth variations in pH all exist (e.g. Fig. 22). Near to the surface these variations are related to temperature changes and biological activity, and at depth to pressure and water-mass transport.

The result is that seawater average pH is widely variable, typically from 7.5 at depth to around 8.5 in surface waters. Surface ocean water pH has also been measured within a similar range[135].

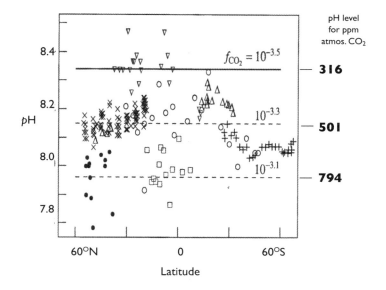

Fig. 22 Measured pH levels for surface ocean waters between 60° N and 60° S in the western Pacific Ocean (after Skirrow, 1965). Note variability between 7.8 and 8.3, and compare that natural variability with the projections of hypothetical mean ocean pH level under atmospheric conditions with 316, 501 and 794 ppm carbon dioxide (calculated by Tom Segalstad after Bethke, 1996 for an idealised ocean model for CO_2 in equilibrium with seawater, without including the effects of solid mineral buffers such as $CaCO_3$). Even with such unrealistic assumptions, the pH change is minor and remains well within modern natural variation.

Finally, pH varies not only with depth and geography, but also with time, over both longer (centennial) and shorter (diurnal) timescales, as a result of variations in oceanography, temperature and biological activity. For example, Revelle & Fairbridge (1957)[136] recorded a pH of 9.4 in isolated coral reef pools during the warmth of the day and 7.5 at night, the reduction being attributable to a lack of photosynthetic uptake but continued respiration; meanwhile, adjacent reef waters remained more or less steady at pH 8.2. Over longer timescales, and using boron isotope analysis, Pelejero et al. (2005)[137] reported variability in pH of 7.9 to 8.2 over a 300 year period using samples collected from a fossil reef coral (*Porites*) on the Great Barrier Reef; and Liu *et al.* (2009)[138] recorded pH fluctuations between 7.9 and 8.3 from Holocene fossil corals from the South China Sea that date back to 7,000 years old. Again, however, the caveat about the reliability of boron isotope analysis must be kept in mind[134].

Henry's Law

The account just given of carbon dioxide saturation is theoretically sound, but nonetheless it can be misleading to speak of the 'saturation level' of carbon dioxide in the ocean.

From first principles, the concentration of carbon dioxide dissolved in seawater is dependent on the partial pressure of carbon dioxide in the atmosphere above the water. This relationship is summarized by Henry's Law[139]. Accordingly, if the ppm level (partial pressure) of atmospheric carbon dioxide is increased, then more of the gas will dissolve in the surface ocean in amounts that are governed by Henry's Law Constant. At the Earth's average surface temperature, the result is that about fifty times more carbon dioxide dissolves in the ocean than is present in the atmosphere. The process is similar to that involved in the manufacture of a bottle of soft drink – the manufacturers increase the pressure of carbon dioxide over the fluid in the bottle, with the result that more carbon dioxide becomes dissolved in the soda water.

If enough carbon dioxide is dissolved in the ocean (and depending too upon the precise chemical balance of carbonic acid, carbonate and bicarbonate ions, and important elements like calcium and magnesium), a level may be reached where the water becomes saturated with respect to solid carbonate minerals like aragonite (calcium carbonate, $CaCO_3$). These minerals may then start to crystallize and precipitate from the solution. Such processes may proceed slowly, faster or instantaneously, depending on the chemical state of the system, and are most common in warm and shallow tropical ocean waters.

The upshot of all this is that as more carbon dioxide is added to the atmosphere, more is in turn dissolved in the ocean and this eventually leads to the accumulation of solid carbonates on the sea floor, either by the direct precipitation in shallow water of aragonite ($CaCO_3$), or, across wide areas of the deep ocean seabed, by the accumulation of the dead calcitic ($CaCO_3$) skeletons of marine unicellular *foraminifera* or *coccolithophorids* that have settled out of the plankton. Therefore, there is no limit, or 'saturation level', as to how much carbon dioxide can be added to such a system; rather, the only limit is the availability of carbon dioxide in the first place.

Considering fossil fuels (coal, petroleum, natural gas) as the source of carbon dioxide that is the focus of public discussion, geochemist Tom Segalstad (1996)[74] has calculated that burning all the currently available fossil fuel resources would increase atmospheric carbon dioxide concentration by just 20 per cent. The rest of the carbon dioxide would quickly and effectively be dissolved in the ocean, to be precipitated as marine calcium carbonate as governed by Henry's Law. It follows from this that it is not possible to double the atmospheric carbon dioxide level by the burning of fossil fuels alone.

Statement of the 'acidification' scare
Against this long background history of ocean buffering and natural variability in pH, and the additional, self-evident fact that modern marine organisms thrive across a wide alkalinity range, deterministic computer modelling is used to imply or assert that:
(i) A meaningful ocean average pH exists at which all, or most, marine organisms thrive.
(ii) This average pH has become less alkaline by 0.075 pH units since preindustrial time (from 8.179 to 8.104 between 1751 and 1994; Orr *et al.*, 2005)[140].
(iii) A doubling of atmospheric carbon dioxide will cause a dangerous further change in ocean alkalinity that is beyond the range of adaptive response of many marine organisms (e.g. Guinotte *et al.*, 2003[141]).

107

The first mentions of 'acidification of the ocean' are found in papers published in the 1990s, but it was a 2005 Royal Society report[142], from a committee chaired by Scottish biologist John Raven, that produced the notorious computer-model prediction that 'the average pH of the oceans will fall by up to 0.5 units by 2100 if global emissions continue to rise at present rates'. The manifold inadequacy of the modelling that had been performed was immediately pointed out by David Bellamy and Jack Barrett[143], and included: that inflated, unrealistic rates had been assumed for fossil-fuel burning; that the model failed to account for the ocean's buffering characteristics, its capacity to sustain large phytoplankton blooms, the river inflows that it receives, and the mechanisms of carbonate deposition; and that the model ignored too the effects of increasing temperature on carbonate saturation. Nonetheless, the acidification scare was off and running, and many other scientists subsequently seized on the matter as a basis for their (now assured-to-be-funded) research. That the scare has since grown longer legs is partly a result of the generous targeted research funds provided in response to its existence. It also reflects media attitudes and the public's consequent lack of defence against advocacy science more than it does any merit in the hypothesis, as detailed further below.

Meanwhile, in the real world…

Carbon dioxide levels in the atmosphere are controlled partly by outgassing from the oceans, and in turn the alkalinity of the surface ocean is generally consistent with the carbon dioxide content of the atmosphere. But both the oceans and the atmosphere are in constant turbulent motion, with consequent variability in gas interchange at different locations, and modulated also by differing time constants up to a millennia long. Ocean pH is therefore not a function of atmospheric carbon dioxide alone, but depends just as much on complex oceanographic and biogeochemical processes, attendant physical changes in temperature, salinity and nutrients, and submarine volcanic sources – a major consequence of which is that ocean surface pH varies widely from one location to another, from less 7.8 in regions of upwelling deep water to about 8.5 in areas of surface sinking[135].

Against this backgound, the persistent, and determinedly alarmist model projections of 'acidification' of the oceans suffer from many inadequacies[144]. These include:

- Inadequate knowledge of the natural flux of carbon dioxide into the ocean, especially from volcanic sources[80], against the contextual background of a possible human input of up to 7 Gt C/yr (Gigatonne of carbon per year) into an ocean reservoir that already contains 39,000 Gt C.

- Inadequate numbers, and inadequate historical, geographical and depth coverage, of measurements of ocean pH values. The available data do not permit accurate calculation of changed pH, and all estimates of its change are unvalidated model studies.

- The inaccuracy of some historical pH measurements, including inadequate correction for the temperature-dependency of pH. Seawater pH is very sensitive to temperature, and it is far from certain that earlier temperature measurements are of sufficient accuracy to constrain the pH value.

- The equilibrium value of atmospheric and dissolved carbon dioxide is very sensitive to the chemistry of ocean surface water. While it is true that a large change in atmospheric carbon dioxide concentration would cause a small and slow change in surface ocean pH, it is equally the case that a small change in ocean chemistry (perhaps caused by upwelling) would provide a more immediate change to atmospheric carbon dioxide concentration. The available historical data are not determinative of which of these processes has dominated, or to what degree, so it remains possible that a variation in surface ocean pH has to a greater or lesser extent contributed to the observed rise in atmospheric carbon dioxide during the twentieth century in line with Henry's Law – whereby any warming of the oceans results in the outgassing of carbon dioxide, concomitantly reducing the amount of dissolved gas and increasing the alkalinity of ocean surface waters.

- The ignoring of experimental data, which show that adding carbon dioxide or iron to the ocean can cause more photosynthesis (fertiliser effect) rather than pH change[145]. This result is consistent with SeaWIFS[146] (Sea-viewing Wide-field-of-view Sensor) satellite measurements, which indicate an increase in oceanic chlorophyll-a levels over the last 15 years[147]. Much of this enhanced productivity encompasses the growth of more plankton with carbonate tests, such as *coccoliths* and planktic *foraminifera*, the dead skeletons of which settle to the seafloor to accumulate as limestone or to be dissolved in deep, unsaturated water.

- The promulgation of alarm about skeletal thinning in planktonic micro-organisms in response to reduced alkalinity[148]. Meanwhile new experimental results which show that the common phytoplankton of the North Atlantic actually grow in size and volume with increasing carbon dioxide are ignored. For example, Iglesias-Rodriguez and colleagues comment that 'field evidence from the deep ocean is consistent with these laboratory conclusions, indicating that over the past 220 years there has been a 40 per cent increase in average *coccolith* mass[149].' (Furthermore, these results directly contradict those of earlier experiments that incorrectly used the direct addition of hydrochloric acid to reduce the pH of experimental ocean water, instead of letting carbon dioxide bubble through)[150].

- A failure to account for an effect of weathering caused by enhanced atmospheric carbon dioxide, whereby the slight acidity of rain (typically pH 5.7[151]) causes an acceleration in chemical weathering, thus driving more calcium down the rivers to encourage the eventual formation of more oceanic carbonates, including limestone, thereby sequestering carbon dioxide below the seabed.

- The false assumption that any decline in pH that is observed should be attributed to increased atmospheric carbon dioxide, whereas some such changes are certain to be related to natural multi-decadal cycles such as the Atlantic Multi-Decadal

Oscillation (AMO) and the Pacific Decadal Oscillation (PDO) which are known to change oceanic temperature and biological productivity; other changes stem from the upwelling of cold, deep water of lower pH.

- The failure to allow for variations in depth of the ocean *lysocline*, the 3.5-4.5 km depth below which calcium carbonate dissolution accelerates; the *lysocline* varies by up to 1,000 m across the ocean basins, and also by about 200 m between glacial and interglacial time. The pH range of the modern ocean is maintained by a fluctuating and self-adjusting balance between calcium carbonate production by lime-secreting organisms (especially plankton), and its accumulation as seabed limestone at depths above the *lysocline*, typically at rates of 1-2 cm/1000 years in deep water and at much greater rates in shallow water; and the destruction of carbonate by dissolution below the *lysocline*.

- A lack of allowance for the proven adaptability of marine organisms and that 'biological processes can provide homeostasis against changes in pH in bulk waters of the range predicted during the 21st century' (Hendriks *et al.*, 2010)[152]; in addition, ancient coral reefs first evolved at a time when the atmosphere contained many times the present concentration of carbon dioxide (cf., Fig. 14), and modern coral reefs are known to thrive in carbon dioxide-rich environments (Pichler and Dix, 1996; Pichler *et al.*, 2000)[152].

Experimental work[153] has demonstrated that additions of HCO_3^- to synthetic seawater increases the calcification rate of reef-building corals over a range up to (for *Porites*) and beyond (for *Acropora*) a concentration of three times that of today's seawater. At the same time, photosynthetic rates in the coral's symbiotic algae are also enhanced up to a saturation point at levels two to three times seawater. As Herfort *et al.* conclude, the likely result of human emissions will therefore be 'about a 15% increase in oceanic HCO_3^-, which 'could stimulate photosynthesis and calcification in a wide variety of hermatypic corals,' thereby sequestering human carbon dioxide emissions into the seabed as limestone.

- A failure to consider the context that the small changes in equilibrium ocean pH projected by chemical modelling, even should they eventuate, lie well within the current natural range of pH variability to which many marine organisms are already comfortably adapted (Fig. 22). For example, experimental work with sea urchin larvae has shown that their development is completely insensitive to variations in pH in the range 6.6-8.2[154]. And that the range of adaptation of marine organisms includes actual acidic, as opposed to slightly less alkaline, conditions is shown by the fact that benthic calcifying organisms like the mussel *Bathymodiolus brevior* live happily near volcanic hydrothermal vents in an ambient pH as low as 5.36[155]. This is but one, albeit spectacular, example of the important fact that marine organisms control their calcifying mechanisms behind organic membranes, which shield carbonate hard parts from direct contact with ocean water.

The death of coral reefs

A particularly persistent claim of the ocean acidifiers is that carbon dioxide-linked changes in ocean temperature and alkalinity will cause mass mortality of coral reefs. Proponents of this claim have recently taken to citing the study by De'ath *et al.* (2009), which purported to document a recent 14 per cent decline in the rate of calcification in the coral *Porites* from the Great Barrier Reef (GBR). It apparently escaped *Science's* referees that this conclusion conflicts with earlier research that showed a statistically significant increase of 4 per cent in GBR coral growth rates during the warmings of the twentieth century (Lough & Barnes, 2000)[156].

Stimulated to enquire into this, Ridd *et al.* (2009) attempted to replicate the results of De'ath *et al.* (2009). They found that a sudden change in calcification indicated in 1990 coincides with a discontinuity in the data, with most cores ending around 1990 and the cores that were sampled after 1990 being taken from different places[157]. Other problems included that the inference of slowed coral growth rate 'depended unduly on questionable end-of-timeseries data from cores that ended in the years 2004/2005,' and that the De'ath *et al.* analysis 'did not allow for the naturally occurring ontogenetic reduction in calcification rate that is evident in the dataset.'

The dataset analysed by De'ath *et al.* does not therefore support the hypothesis that coral calcification has slowed recently on the GBR, whether due to acidification of the ocean or for any other environmental reason. Furthermore, both laboratory experiments[158] and measurements on natural coral reefs[152] demonstrate that corals thrive in carbon dioxide-rich environments.

Conclusion

Both the sea-level and the ocean acidification scares are environmental exaggerations for which, to adopt Nigel Lawson's redolent phrase, a grain of truth has been embedded in a mountain of nonsense.

What the IPCC thinks will happen to the notional global sea-level has little bearing on the real-world coastal engineering problems that should be the concern of sovereign governments. And the concept of 'ocean acidification' is based on computer modelling of the reaction of adding carbon dioxide to water in an unreal ocean without organisms, dissolved solids or mineral and rock buffers – and projecting that a large increase in atmospheric carbon dioxide will cause a small reduction in pH. In effect, the pH of surface water in this virtual reality ocean is expected to become slightly less alkaline, and to move closer to the mean pH of the oceans. But at the same time, the pH is projected to remain well within the range of tolerance of most marine organisms. As geochemist Tom Segalstad has pointed out, true acidification of the world ocean would require average pH to be reduced below six to be able to dissolve solid carbonates, an impossible scenario given the ocean's effectively limitless buffering capacity[144].

The idea that a continued small rise in average global sea-level, or a small shift in ocean average pH, should either occur, will cause catastrophic environmental damage is fundamentally implausible. And the more so because, as for forward projections of atmospheric temperature, the alarm in both these cases again centres around the projections of unvalidated computer models rather than being rooted in empirical evidence.

Coda

> 'Do you believe in global warming?' the reporter asks again (meaning, of course, 'Do you believe in dangerous global warming caused by human carbon dioxide emissions?')

> 'It depends,' I reply. 'For there are many different realities of climate change.'

In Chapter 5, we will turn our attention fully to the second of those realities, which we have incidentally started to touch on in this chapter. It is the virtual reality of deterministic computer models.

5 Computer models: projection not prediction

... No simplification of the [GCM] equations can accurately predict the properties of the solutions of the differential equations. The solutions are often unstable. This means that a small variation in initial conditions will lead to large variations some time later. This property makes it impossible to compute solutions over long time periods.

As an expert in the solutions of non-linear differential equations, I can attest to the fact that the more than two-dozen non-linear differential equations in weather models are too difficult for humans to have any idea how to solve accurately. No approximation over long time periods has any chance of accurately predicting global warming....

To base economic policy on the wishful thinking of these so-called scientists is just foolhardy from a mathematical point of view. The leaders of the mathematical community, ensconced in universities flush with global warming dollars, have not adequately explained to the public the above facts.

(Peter Landesman, 2009[159])

Of the three main arguments advanced by the Intergovernmental Panel on Climate Change (IPCC) in 2001 in support of the idea of dangerous human-caused global warming (see p. 55), the only one that is left standing today is that general circulation models (GCMs) project significant, though by no means disastrous, warming by the end of the twenty-first century. For example, results reported in the Fourth Assessment Report (IPCC, 2007) project that the average global warming by the end of this century will be ~2-4°C, based on the IPCC's lowest emissions scenario, B1, which projects that carbon dioxide will attain 600 ppm by century's end (Fig. 23). How seriously should we treat this estimate?

Deterministic computer models are not evidence

There is no Theory of Climate, in the sense that there is a Theory of Gravitation or General Relativity. Therefore no theoretical computer model constructed from first physical principles, including the GCMs that are employed by the IPCC, can accurately predict future global, let alone regional, climate[160]. A rare insider insight into the systematic deficiencies of GCMs by an expert modelling practitioner has been provided recently by former Australian Commonwealth Scientific and Industrial Research Organization (CSIRO) scientist, John Reid[161].

Parameterization

All GCMs use complex partial differential equations to describe the ocean-atmosphere climate system mathematically. When fed with appropriate initial data, these models can calculate possible future climate states. The models are deterministic, in that every factor that is known to influence climate significantly must be included at the start in order to allow the model to correctly determine a future climate state. Deterministic computer models therefore assume (wrongly) that we have a complete understanding of the climate system. Since some climate processes are too complex or small in scale to be fully physically represented in the model, they are replaced by simplified *parameterized* (read, educated-guess) relationships. For example, nearly all GCMs include parameterizations for atmospheric convection, land surface processes such as reflectivity and hydrology, and cloud cover and its microphysics.

Parameterized processes that are known to play a particularly vital role in climate regulation include cloud formation, deep tropical convection, and mid-latitude storm creation. The uncertainty regarding these processes is often very high. For example, depending upon the parameterization adopted for clouds, climate sensitivity for a doubling of carbon dioxide ranges between 1.5°C and 11.5°C[162], and a recent comprehensive study has demonstrated that major imperfections exist in cloud simulation by all ten atmospheric GCMs that were tested[163]. Parameters are susceptible to tweaking and the result is that the output of a heavily parameterized model inevitably ends up conforming with its creator's expectation. It is no surprise that virtually all published GCMs predict rises in temperature under increasing greenhouse gas forcing, because that is precisely what they are designed to do.

Validation

Climate model projections are only meaningful to the extent that they accurately mimic reality. This should be tested by checking the models against real-world data in a process termed validation. In the sense used by computer engineers, validation requires that a model be rigorously tested to demonstrate that it can forecast the future behaviour of the modelled system to a specified, satisfactory level of accuracy. No such procedure is known to have been carried out for any of the climate models used by the IPCC. Further, it appears that no recent IPCC document, including the Fourth Assessment Report, discusses the validation of their GCM models, and neither does the word 'validation' appear in the Glossary of the 2007 Fourth report.

The Kelvin Fallacy

Because they presume that the physics of the climate system is fully understood, GCMs conform to the Kelvin Fallacy[164], i.e. they presume that there are no 'unknown unknowns' as Dick Cheney so elegantly put it. But climate is far from being fully understood. We have just seen above that the effect of clouds is one poorly constrained parameter, the net warming feedback caused by increased atmospheric carbon dioxide is another (climate sensitivity; see Chapter 3), and there are many more.

Despite these fundamental limitations, there is a widespread public misapprehension, which the IPCC and their modellers do nothing to correct, that the forward climate projections of computer GCMs can provide predictions of future climate. At the same time, GCM hindcasts, that match an historic temperature record, are trumpeted as 'evidence' for human-caused warming. In fact, such hindcasts are achieved by selectively adjusting various tuneable program parameters within plausible ranges until a target curve (usually the historical temperature record for the last 150 years) is more or less successfully matched. Furthermore, in concentrating on hindcasting one climatic parameter alone (temperature), the modellers are ignoring other important aspects of climate that are much more difficult to hindcast, especially in the ensemble – such as precipitation, hurricanes and snow storms.

In other words, GCM forward projections are not climate 'forecasts' and neither are GCM hindcasts 'evidence'. The projections have no demonstrated statistical skill, and instead represent virtual reality futures selected out of many alternatives of equal modelling merit. Conversely, GCM hindcasts that match an historic temperature curve are exercises in curve-fitting – to help with which a typical GCM has a very large number of degrees of freedom, running into millions. As Johnny von Neumann famously remarked about curve fitting: '...with four parameters I can fit an elephant, and with five I can make him wiggle his trunk.'

This simple insight appears to have escaped the modellers, who continue to claim vindication of their models because they can provide output that retrofits the 'known' twentieth century temperature record (IPCC, 2001; TAR, Chapter 12, Fig. 12.7). In von Neumann's terms their elegant climate elephant is somersaulting with a double pike, but its performance is nonetheless an inevitable result of the modelling techniques used and the numerical values assigned to key variable parameters.

Should the recently announced British Meteorological Office review of the Hadley Climatic Research Unit (CRU) temperature record result in significant changes to the estimated temperature curve for the last 150 years, watch for a scurrying among the IPCC modelling teams to provide new outputs that mimic the new curve: 'We wuz wrong before,' they will say, 'but trust us, we've got it right this time.'

Other limitations of GCMs

Soon and colleagues (2001)[165] and Essex (2008)[166] have provided comprehensive discussions of a wide range of the deficiencies that GCMs have when used as predictive, rather than heuristic, tools. When the current generation of models are tested against factual data, those deficiencies include the following.

- No IPCC model was able to successfully forecast the temperature record that actually elapsed between 1990 and 2009. What happened in reality was that a rising temperature cycle peaked in 1998 and declined thereafter. Instead, all models projected the ensemble occurrence of monotonic warming. Further, projection of the current measured cooling trend indicates that global temperature is now tracking outside the low estimate error boundary of the IPCC suite of model projections (Fig. 23)[167].

- This IPCC forecast failure is scarcely surprising, because the cyclicity represented in the real world climate data is probably both solar-forced and also related to internal climatic phenomena such as the North Atlantic Oscillation (NAO), the Atlantic Multi-Decadal Oscillation (AMO), the Pacific Decadal Oscillation (PDO) and the El Nino-Southern Oscillation (ENSO); until very recently[168] none of these vital, climate-influencing mechanisms were included in the GCMs.

- The models are unable to simulate the distribution of temperature up through the troposphere; the upper

Global Mean Temperature

From: ICC AR4 report Figure 10.26 on p. 803 of Chapter 10 of the Working Group I (AIB)

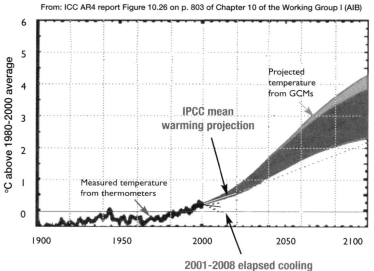

Fig. 23 Comparison between measured surface temperature to 2000 (black line), IPCC model projections of future temperature (white line plus scatter of estimates represented by grey envelope) and projection of the 2001-2008 cooling trend and dotted error bounds. Note that all IPCC projections now fall outside the error bounds of the elapsed temperature record.

troposphere 'hotspot' that they predict is not observed in meteorological measurements (Fig. 24, inside back cover)[13], which indicates that the alleged positive temperature feedback processes are not as strong as assumed by the GCMs[169]. This is because the models overestimate climate sensitivity, as judged by satellite measurements of outgoing infrared radiation. Models assume a feedback factor of three, whereas the missing hotspot limits the feedback factor to no more than 1.2, and other empirical tests (e.g. satellite data, cloud observations) suggest a feedback factor of 0.5 or less[87], which in turn implies that the actual temperature rise due to the forcings considered by the IPCC can be at most 20 per cent of what the IPCC predicts.

- The models underestimate surface evaporation in response to increased temperature by a factor of three. This is a particularly troublesome failure given the importance of water vapour as a greenhouse gas[170], and exaggerates the surface temperature response to solar forcing while underestimating the global precipitation response[126].

- Recently, Koutsoyiannis et al.[171] compared 18 years of projections by global warming models against real-world rainfall and temperature from eight geographically dispersed stations that each has a record over 100 years. It transpires that the more recent IPCC Fourth Assessment Report (2007) model results are no better than older Third Assessment Report (2001) versions, and that overall the 'models perform poorly, even at a climatic (30-year) scale. Thus local model projections cannot be credible, whereas a common argument that models can perform better at larger spatial scales is unsupported.'

- Models incorrectly project a continuing increase in ocean heat content, whereas observations demonstrate no increase for the last five years.

Ocean temperatures have only been measured adequately since the beginning of 2004, using the Argo network of more than 3,000 drifting ocean probes that are deployed worldwide. The probes dive from the surface to depths up to 2 km, resurfacing every ten days to transmit their measurements to scientists on shore via satellite, allowing salinity and density depth profiles to be calculated. Earlier measurements, made during the twentieth century from ships of opportunity, were not collected under controlled conditions and are therefore of dubious quality and usefulness[172]. The Argo data currently show slight cooling since 2004[173]. Loehle[174] notes that the Argo-measured cooling trend is also reflected in falling upper ocean heat content, and DiPuccio[175] points out that ocean temperatures actually need to be rising above a certain rate to be consistent with the IPCC greenhouse warming hypothesis.

The fact that they aren't, and that there is nowhere for the alleged heat to be hiding, amounts to another negative test of the dangerous warming hypothesis.

• GCM calibration is faulty. Although human emissions weren't large before 1940, the models assume that the temperature rise since 1850 is due to the currently considered forcings, principally human carbon dioxide. In fact, since 1850 the Earth's temperature has been in a phase of recovery from the Little Ice Age; the calibration of the models assumes that this natural warming trend is instead almost entirely caused by carbon dioxide emissions, and that in full knowledge also that ice core data show that temperature changes precede carbon dioxide changes during natural climatic cycling (Chapter 3).

The numerous mismatches that occur between GCM model projections and factual reality are caused, among other things, by the models' failure to simulate cloud processes accurately, or to incorporate known multi-decadal oscillations (e.g. the AMO and PDO; also, shorter term ENSO cycling) and solar forcings other than direct radiance variation.

Simpler models might be better
GCM models have been developed under the assumption that the more comprehensive they are, the more useful will be their results. This may be incorrect, for simple models that deliberately use a limited number of variables to simulate complex natural processes are, in at least some cases, more accurate than more complex ones. This has been shown, for example, for ENSO cycling by Halide & Ridd (2008)[176], and for the hydrological components of one of the Land Models used to provide information to the IPCC[177]. As the former authors conclude:

> If larger and more complex models do not perform significantly better than an almost trivially simple model, then perhaps future models that use even larger datasets,

and much greater computer power may not lead to significant improvements in both dynamical and statistical models.

GCM *models are not predictive tools*

The most important general point that is all but completely ignored in the public debate was stated at the outset of this section, and cannot be overstressed. It is that GCM models do NOT provide future climate predictions or forecasts. Rather, the models produce 'projections' – which have no demonstrated forecast skill, and are merely selected outputs from among the innumerable alternative climate futures that might or might not eventuate. This has been well summarized by IPCC senior scientist and lead author, Kevin Trenberth, who writes[178]:

> There are no (climate) predictions by the IPCC at all. And there never have been. [Instead, there are only] 'what if' projections of future climate that correspond to certain emissions scenarios...

> [For] none of the models used by IPCC is initialised to the observed state and none of the climate states in the models corresponds even remotely to the current observed climate... [GCMs] do not consider many things like the recovery of the ozone layer, for instance, or observed trends in forcing agents...

> [And] the state of the oceans, sea ice and soil moisture has no relationship to the observed state at any recent time in any of the IPCC models.... There is neither an El Nino sequence nor any Pacific Decadal Oscillation that replicates the recent past; yet these are critical modes of variability that affect Pacific rim countries and beyond... the starting climate state in several of the models may depart significantly from the real climate owing to model

> errors (and) regional climate change is impossible to deal
> with properly unless the models are initialised.

That deterministic GCMs are unable to predict future climate accurately, at both global and regional level, is not just a matter of Kevin Trenberth's opinion but is well understood by all climate modelling practitioners and their colleagues, as summarized by the IPCC authors who wrote; 'In climate research and modelling, we should recognize that we are dealing with a coupled non-linear chaotic system, and therefore that long-term prediction of future climate states is not possible' (IPCC Third Assessment Report, p.774)[5].

In a very real sense, then, GCMs are scientific computer games. Like many other types of scientific computer model, their value is heuristic, in which capacity they are valuable learning and discovery tools. Their practitioners understand full well that GCMs cannot provide accurate predictions of future climate, which is precisely why they use the term 'projection' to describe their outputs.

Alas, the vital distinction between a projection and a prediction is all but completely lost on (or wilfully misreported by) journalists and media producers, and therefore also on politicians and the general public. As the IPCC understands only too well, the computer-based, speculations that it makes about future climate amount to climate prediction so far as the general populace is concerned.

Statistical climate modelling

Though it seems almost to be a secret, at least so far as public knowledge is concerned, there is a scientifically well understood, alternative way in which computers can be used to project future climate. This method constructs projections by assembling data on past climate change, identifying patterns within them, and then projecting these patterns into the future. Such models are statistical and empirical. They make no presumptions about complete understanding; instead, they seek to recognise and project into the future the climate patterns that actually exist in past real world data.

In 2001, in an early application of this type of modelling, Russian Sergey Kotov[179] used the mathematics of chaos to analyse the atmospheric temperature record of the last 4,000 years from a Greenland ice core. Based on the pattern that he recognized in the data, Kotov extrapolated cooling from 2000 out to about 2030, followed by warming up to the end of the century and 300 years of cooling again thereafter.

In 2003, Russian scientists Klyashtorin and Lyubushin[180] analyzed the global surface thermometer temperature record from 1860 to 2000, and identified in it a recurring 60-year cyclicity. This multi-decadal cycle probably relates to the Pacific Decadal Oscillation (PDO), which can be caricatured as a large scale El Nino/La Nina climatic oscillation[181]. The late twentieth century warming that is of such concern to the IPCC corresponds to the most recent warm half-cycle of the PDO, which has now reversed its phase and projects forwards as global cooling of 0.2°C and Arctic cooling of 1.5°C out to 2035 (Fig. 25). Recently, it has been confirmed by US ecologist Craig Loehle that the 2003 Klyashtorin projection matches closely with the measured satellite microwave sensing units (MSU) data for 2003-2009[64].

Choosing the somewhat longer period of the last 1,000 years over which to establish the historic temperature pattern, in 2004 Craig Loehle[182] used simple periodic models to analyse climate records of sea surface temperature from a Caribbean marine core and cave air temperature from a South African stalactite. Without using input data for the twentieth century, six of his seven models showed a warming trend similar to that in the instrumental record over the past 150 years; and projecting forward from that point, the best fit model foreshadows cooling of between 0.7 and 1.0°C during the next 20-40 years.

In 2007, the 60-year climate cycle was identified again, this time by Chinese scientists Lin Zhen-Shan and Sun Xian[183], who

ΔT, °C

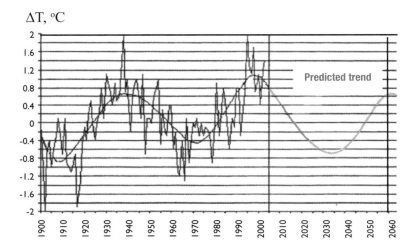

Fig. 25 Annual average air surface temperature for the Arctic region for 1900-2006, with fitted 60-year curve that delineates the underlying multi-decadal cyclicity projected forward to 2060. Note that the current phase of cooling is projected to last until 2035.

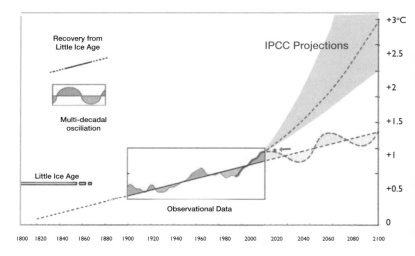

Fig. 26 Reconstructed surface temperature curve by Akosofu (2009), which shows the warming recovery from the Little Ice Age, on which are superimposed multi-decadal oscillations (MDO). In contrast to the IPCC computer model projections (dotted central line and grey envelope, after 2008), actual temperature records are at present tracking along the MDO-cooling projected on this graph (small black dot, arrowed).

used a novel multivariate analysis of the 1881-2002 temperature records for China. These authors showed that temperature variation in China leads parallel variation in the global temperature by five to ten years, and has been falling since 2001. They conclude, 'we see clearly that global and northern hemisphere temperature will drop on century scale in the next 20 years.'

Finally, Syun Akasofu[56] of the International Arctic Research Center has drawn attention to the importance of the same 60 year-long, intermediate scale cycles and pointed out that they are superimposed upon a general warming recovery since the depths of the Little Ice Age (LIA). Most recently, these cycles have been manifest as cooling for 1882-1910 and 1944-1975, and warming for 1910-1944 and 1975-2001. Projecting this pattern into the future suggests that the renewed cooling that started in 2001 marks the start of the next cooling cycle, which will continue until around 2030 with warming likely for another 30 years beyond that (Fig. 26). Even when superimposed on an assumed post-LIA warming trend of 0.5°C/century, Akasofu's projected temperature path lies markedly below the IPCC's GCM projections.

To conclude, the seven scientists whose work is summarized above have applied several fundamentally different mathematical techniques to several different datasets to show that their statistical models forecast that climatic cooling – say that again; cooling – will occur during the first decades of the twenty-first century. If that sounds familiar, it is because it is; for we arrived at a similar conclusion in Chapter 2, drawing on evidence from solar cycle studies.

A naive forecast model

Many scientists, including those advising the IPCC, are unaware of the existence of a branch of the social sciences that examines the principles and methods that are needed to achieve high quality forecasting. With its own website (forecastingprinciples.com), international institute, international journal and a set of 127 principles of good practice[184], the development of evidence-based forecasting owes much to the efforts of Professor Scott Armstrong at The Wharton School of Business, University of Pennsylvania, and his colleagues elsewhere.

Two recent forecasting studies by Kesten Green and Scott Armstrong have shown that the IPCC's approach to climate forecasting is badly flawed[185]. The first paper demonstrated that IPCC's forecasting techniques violated 72 out of the 127 agreed principles for good forecasting, in part because the models are unnecessarily complex. This observation led the same authors, together with astrophysicist Willie Soon, to publish a second paper in which a simple, 'naive' model of climate change was assessed.

The naive model assumes that the temperature next year will remain the same as that of the previous year, which might seem a counter-intuitive assumption given observed climate variability. Modelling proceeded by simulating annual forecasts from 1 to 100 years into the future, starting with the average temperature for 1850 and then extending the process for successive years up through 2007. In this fashion, a total of 10,750 annual average temperature forecasts were produced. Comparing these forecasts with the 'reality' of the Hadley Climatic Research Unit (CRU) temperature curve (1850-2008), the naive model displayed an average error 7.7 times smaller than the error in the IPCC's temperature projection. The authors concluded that:

> Global mean temperatures have been remarkably stable over policy relevant horizons. The benchmark forecast is that global mean temperature for each year for the rest of this century will be within 0.5°C of the 2008 figure.
>
> There is little room for improving the accuracy of forecasts from our benchmark model. In fact, it is questionable whether practical benefits could be gained by obtaining perfect forecasts. While the Hadley temperature data drifts upwards over the last century or so, the longer series shows that such trends can occur naturally over long periods before reversing.

Conclusion

Deterministic computer modelling is unable to provide predictions of future climate suitable for application in policy settings. The GCM projections of warming favoured by the IPCC have been shown to have much larger errors than a simple (naive) model, and they also conflict completely with alternative projections made using statistical modelling. In contrast to the GCMs, most statistical models forecast that climatic cooling will occur during the first decades of the twenty-first century. Temperature measurements confirm that a global cooling has indeed been underway since around the turn of the century, though the final length and intensity of this trend remains unknown. The question that has to be posed is: 'Why is this cooling such a surprise to so many persons, and why are so many still trying to deny it?'

An attempt to answer this question forms the concluding chapters of this book. First, however, we will digress briefly in Chapter 6 to discuss scientific methodology, especially as it pertains to circumstantial evidence and hypothesis testing.

6 Circumstantial evidence and the null hypothesis

We are to admit no more causes of natural things than such
as are both true and sufficient to explain their appearances.

(Isaac Newton)

So far, I have made little mention of the bulk of the climate
alarmist material that now fills out our daily newspapers and news
bulletins, which for so long has been asserting shrilly that many
and varied aspects of Earth's natural systems are being destabilised
by human-caused climate change.

We will all be rooned, they say, as will the polar bears and
armadillos, by melting ice, rising sea-level, more or more intense
storms, more or more intense droughts, more or more intense
floods, more or less precipitation, more or less atmospheric aerosols,
more mosquito bites, more deaths from heat stroke or even – as I
read in an apparently straight-faced newspaper report a little while
back – the collapse of our sewage systems from additional and
excessive rainfall runoff. British engineer John Brignell has
assembled a mighty list of guffaws of 690 of these rhetorical
sillinesses[49]. The list – which starts with acne, progresses through
circumcision in decline, haggis threatened, polar bears deaf, seals
mating more, short-nosed dogs endangered, and finishes with
yellow fever – is beyond parody.

Though it is mother's milk to a geologist, it has mostly escaped
media reporters and other influential public commentators on
climate change, such as Al Gore, that the Earth is a dynamic
planet. Earth's systems are constantly changing, and its lithosphere,
biosphere, atmosphere and oceans incorporate many complex,

homoeostatic, buffering mechanisms. Changes occur in all aspects of local climate, all the time and all over the world. Geological records show that climate also changes continually through deep time. Change is what climate does, and the ecologies of the natural world change concomitantly, in response.

This change is manifest in the daily media reports that we receive about various aspects of Earth's natural systems, such as changes in atmospheric composition, atmospheric aerosol load, global and regional ice volume, the frequency and intensity of storms, patterns of precipitation and drought, sea-level and the range or abundance of individual organisms and their ecological habitats. The press, compliantly repeating what is fed to them by green lobbyists, assert that these changes are controlled by, or linked to, human-caused global warming, mostly without a trace of critical analysis.

All these matters are, of course, proper topics for concern, and in turn many are being subjected to intensive research. But no research study has yet established a certain link between changes in any of these things and human-caused global warming.

Gore's film, 'An Inconvenient Truth'

No-one has done more to mislead the public about global warming, using deliberately manipulated circumstantial evidence, than the former vice-president of the United States, Al Gore. In 2006, this famous giga-green launched his supposed documentary film, *An Inconvenient Truth*, drawing large audiences and eulogistic reviews. The film is a masterpiece of global warming evangelism, and uses every artifice in the propaganda filmmaker's book. Dramatic and beautiful images of imagined climate-related natural disasters segue fluidly one into another: from collapsing ice sheets to shrinking mountain glaciers, from giant storms and floods to searing deserts, and from ocean current and sea-level changes to drowning polar bears. Never explained is the minor detail that all of these events reflect the fact that we humans inhabit a dynamic planet. Certainly, all the environmental changes featured in the film have occurred naturally many times in the past, long before human activities could possibly have been their cause.

When asked 'Do you scare people or give them hope?' in an interview[186] with *Grist Magazine's* David Roberts, Gore replied:

> I think the answer to that depends on where your audience's head is. In the United States of America, unfortunately we still live in a bubble of unreality. And the Category 5 denial is an enormous obstacle to any discussion of solutions. Nobody is interested in solutions if they don't think there's a problem. Given that starting point, I believe it is appropriate to have an overrepresentation of factual solutions on how dangerous it [global warming] is, as a predicate for opening up the audience to listen to what the solutions are, and how hopeful it is that we are going to solve this crisis.

Gore's dishonest efforts to influence the public debate on global warming did not stop at his film. At the same time, he circumnavigated the world giving an expensive lecture to high-profile audiences, and quite unashamedly set up training boot-camps at which global warming devotees were coached in the delivery of his alarmist lecture. At one such camp in Nashville, Tennessee 'more than 1,000 individuals [were trained] to give a version of his presentation on the effects of – and solutions for – global warming, to community groups throughout US.' Not content with one continent, Gore has also run camps in Australia more than once, the first time training 75 volunteer 'climate changers' to replicate the PowerPoint presentation on which *An Inconvenient Truth* was based. Each volunteer was required to guarantee to deliver at least ten seminars over the following 12 months, for a minimum total of 750 sessions across Australia. This propaganda exercise was funded by local environmental lobby group, the Australian Conservation Foundation.

Meanwhile, on the other side of the Atlantic, in late 2006 a British parent and school governor discovered that the government was going to provide all secondary schools in the UK with a video copy of *An Inconvenient Truth* for use in the classroom[187]. Taking exception to such blatant indoctrination of his children about a political matter, Stewart Dimmock took the secretary of state for education to the High Court in London, seeking an injunction for Gore's film package to be withdrawn from the schools. Though the Court declined to recall the film, in a famous victory for commonsense[188] Justice Michael Burton commented[189] that 'the claimant substantially won this case,' and ruled that the science in the film was used 'to make a political statement and to support a political programme' and contained nine fundamental errors of fact (out of 35 that were listed by Mr Dimmock's scientific advisors[190]). Justice Burton required that these errors be summarized in new guidance notes that were to be used as an accompaniment to future educational showings.

The nine errors identified in *An Inconvenient Truth* by Justice Burton consisted of erroneous or exaggerated statements about sea-level rise, evacuation of Pacific Islands, intensity of ocean current circulation, cause/effect relationship between increasing carbon dioxide and increasing temperature, melting glaciers on Mt. Kilimanjaro, the drying up of Lake Chad, the cause of Hurricane Katrina, polar bears and coral reefs. This comprises a list of most of the pin-up environmental scares that are used by global warming alarmists worldwide in support of their cause. In short, the London High Court judgement firmly typed Gore, and the environmental organizations that work with him to spread global warming alarm, as evangelistic proselytisers for an environmental cause, and as abusers rather than users of scientific information.

Over-reliance on circumstantial argument

The various lines of 'evidence' for climate change adduced by Gore, the Intergovernmental Panel on Climate Change (IPCC), and many other lobby groups, are not direct evidence for human attribution at all. First, because all are consistent with natural change; and, second, because many of the changes are currently proceeding in the opposite direction of that expected in a warming world. In particular, and despite the widespread alarmism about these features in the media, the number of tropical storms is not increasing, the global ice volume is not decreasing, the rate of sea-level rise is not accelerating and the number of polar bears is not decreasing. An excellent and independent analysis of these and other climate-related scares, with many references, is provided by the 2009 Report of the Nongovernmental Panel on Climate Change (NIPCC)[112].

Correlation does not prove causation

An implicit scientific malpractice that underlies these and many other of the public claims of 'evidence for global warming' is the confusion of correlation and causation. Melbourne climate analyst John McLean is fond of pointing out that the fact that the birds chirrup outside his window every morning as the sun rises is not evidence that the chirruping causes the dawning of the day. Similar logic, or rather the lack of it, applies with great force to two of the most common arguments used in support of dangerous global warming.

The first is the close correlation that is observed in ice cores between temperature change and atmospheric carbon dioxide content. Prior to the availability of high resolution datasets, climate alarmists, ignoring the well-known solution chemistry of carbon dioxide in the ocean (which implied outgassing during warming climate phases; Chapter 3), alleged that this correlation was the 'smoking gun' that proved carbon dioxide villainy. Subsequently, when high-resolution datasets became available, it became apparent that changes in temperature preceded parallel changes in carbon dioxide by at least several hundred years[191], which established firmly that the potential causality is that of temperature change causing atmospheric carbon dioxide change.

As an aside, it was this demonstrated time lag of up to many hundreds of years that gave rise to the appealing idea that the steady rise in carbon dioxide measured during the second half of the twentieth century may represent the oceanic response to the natural warming that followed the last (mid-nineteenth century) cold phase of the Little Ice Age. IPCC scientists believe that such an interpretation is precluded by the evidence of changing atmospheric carbon isotope ratios during the twentieth century, which they interpret as uniquely consistent with fossil fuel burning (IPCC, 2001, Section 3.3.35)[192]. The argument is not that easily resolved, however, for other scientists maintain that the same changes in carbon isotope ratio are in fact consistent with only a small contribution towards rising carbon dioxide levels coming from human sources (Chapter 3, and endnote 97).

The second way in which the correlation-proves-causation argument is endlessly used by warming alarmists involves the many examples of documented changes in the natural world (more polar bears, fewer polar bears; birds nesting earlier, birds nesting later; more droughts, fewer droughts; more floods, fewer floods; more hurricanes, fewer hurricanes; and so on, *ad infinitum*) which are claimed to be accompanied by a rising temperature (often not the case, and especially so if the IPCC's flawed Hadley Climatic Research Unit (CRU) figures are being relied upon), the whole package then being alleged or implied to be 'proof' of the phenomenon of human-caused warming. That such claims are scientifically juvenile does not prevent the media from promulgating them; but in the absence of demonstration otherwise, such changes are simply part and parcel of the way that our dynamic planet goes about its everyday business.

In Australia, phases of drought, bushfire and flood are commonly asserted to be linked to human-caused global warming, especially in the environmentally iconic Murray-Darling river system. Recent research articles[193] have demonstrated that atmospheric circulation systems originating in the Indian Ocean control much of this climatic variability. Widespread claims that recent drought in the Murray-Darling catchment is a result of warming caused by carbon dioxide emissions, for example by the

Australian Bureau of Meteorology[194], rest on a faulty understanding of the physics of evaporation. As demonstrated in a recent paper by Lockart and co-authors[195]; 'During drought, when soil moisture is low, less of the sun's radiant energy goes into evaporation and more goes into heating the atmosphere which causes higher temperatures. Most importantly, the higher temperatures do not increase evaporation but are actually due to a lack of evaporation and this is a natural consequence of drought.'

In both its second and third assessment reports, the IPCC agreed with these conclusions, for example stating in 1996 (p. 173) that 'overall, there is no evidence that extreme weather events, or climate variability, has increased in a global sense, through the twentieth century.' A similar conclusion was reached by the editors of the journal *Natural Hazards*, in a June 2003 issue that focused on assessing whether global warming causes extreme weather.

These environmental issues, and the other iconic 'global warming problems' that are cited by the IPCC and its activist supporters, are never as clear as they are persistently made out to be. None of these changes are evidence for human-caused warming *per se*, many are actually changing in ways that are the opposite of an expected response to warming, and some are not even necessarily associated with rising temperature. For example:

- Sea-level rise is not accelerating; instead, the global average sea-level continues to increase at its long-term rate of 1-2 mm/year globally, at the same time as local and regional sea-levels continue to exhibit typical natural variability – in some places rising and in others falling (Chapter 4).

- Unusual sea-level rise is not drowning Pacific coral islands, and nor are the islands being abandoned by 'climate refugees'; instead, the best available data exhibits dynamic variations in Pacific sea-level that accord with El Nino-La Nina cycles (Chapter 4) and are probably superimposed on a natural, gentle, long-term eustatic rise. Persons emigrating from the islands are doing so for social and economic reasons rather than in response to environmental threat.

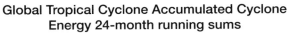

Global Tropical Cyclone Accumulated Cyclone Energy 24-month running sums

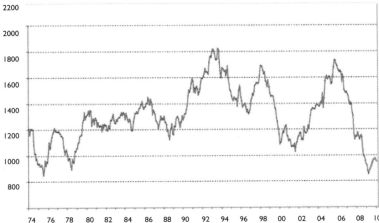

Fig. 27 Summary plot of the total energy contained in the tropical storm systems of the world since 1974 (Ryan Maue). Note the lack of any long-term trend, and that 2005 (Hurricane Katrina) was a year of unusually high activity, since when the level of tropical storm energy has decreased to a historic low.

- The intensity and number of tropical storms are not increasing. Over the short term, 2009 represents a low point in the global energy index of tropical storm power in the last 35 years (Fig. 27) (Maue, 2009 a, b)[196]. Reliable global figures are not available over longer time-scales, but Australian scientist Jonathan Nott and his colleagues have shown that in tropical Queensland more land-falling cyclones occurred during the fifteenth, seventeenth and nineteenth centuries (Little Ice Age) than during the warmer twentieth century, and that the incidence of cyclones during the twentieth century was less than the long-term average rate of occurrence over the last 6,000 years[197].

- Hurricane Katrina was not caused by global warming[198]; rather, it represented a typical, natural tropical storm, if somewhat stronger than most.

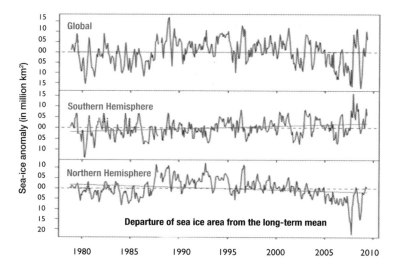

Fig. 28 Sea ice anomalies 1979-2009 (May) globally (top), southern hemisphere (middle) and northern hemisphere (bottom). Note that (i) the post-1998 trend of decreasing sea-ice area in the Arctic has reversed over recent years; and (ii) sea-ice around Antarctica has recently been significantly above the long-term average. The global sea-ice area shows no trend over the measured period.

- Following on from the two previous points, the increase in storm damage insurance claims since the 1990s has nothing to do with global warming; instead, it results from the increasing number of people who choose to live in attractive but hazard-vulnerable coastal regions[199].

- The Arctic region was warmer in the early 1940s than in the 1990s, and has in fact actually cooled since 1920[200]. Scientists at the leading Russian Arctic and Antarctic Research Institute have repeatedly stressed the evidence for a quasi 60 year-long global climatic cycle, the Arctic manifestation of which turned down in 2000-2003 and is projected to result in decreased air temperatures and increased northern ocean sea-ice cover for the next 10-20 years (Fig. 25)[201]. Przybylak (2000) has shown that the late twentieth century warmth in Greenland was similar to that observed in the nineteenth

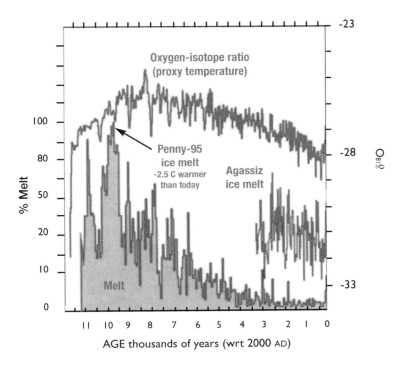

Fig. 29 Reconstructed sea-ice cover for the Eastern Arctic ocean from two ice core records (Agassiz and Penny). Note the virtual absence of sea-ice during the early Holocene climatic optimum, 9,000-10,000 years BP, and the steady increase in ice cover since then – consistent with the long-term cooling trend delineated by the oxygen isotope record (upper curve). Even during the recent 2007 melting episode (Fig. 28), the amount of sea-ice in the Arctic Ocean remained much greater than the long-term Holocene average.

century and that the 'highest temperatures since the beginning of instrumental observation occurred clearly in the 1930s,' the latter result being confirmed by Chylek *et al.* (2004)[202].

- Melting rates of the polar ice caps are not increasing dangerously, and nor do they fall outside their likely natural rates; instead, both Greenland and Antarctica are cooling[203], and accreting ice on their interior (at rates of 6.4 cm/yr and 0.6 cm/yr, respectively) and melting around their periphery, and are close to mass balance[204]; outlier claims of ice-mass loss for Antarctica[205] use gravity data from the GRACE satellite which require complex modelling treatment for use in estimating ice mass. Comiso (2000) showed a cooling rate for Antarctica of 0.08°C/decade from weather station data and 0.42°C from satellite observations, and Doran *et al.* (2002) recorded a 0.7°/decade cooling for the McMurdo Sound Dry Valley region in Antarctica between 1986 and 2000.

- Regarding sea-level change[206], Zwally *et al.* (2005) provide the needed perspective, as they record recent rates of equivalent sea-level rise from combined ice cap melting of 0.05 mm/year, which projects as a trivial rise of 5 cm over the next one thousand years. Indeed, modelling by Krinner *et al.* (2007) projects that Antarctic ice accretion that will lead to a *decrease* in sea-level by 2100 of 1.2 mm/year, with a cumulative drop by then of 6 cm. Meanwhile, in 2009, sea ice around Antarctica lay close to the long-term, 1979-2008, average (3.14 compared with 3.23 million sq km[207]), and though Arctic sea ice area was 0.86 million sq km below average, in the past the Arctic Ocean has had much less ice than today, including none at all (Fig. 29)[208].

- Ice sheet models that project dangerous rates of disintegration of the polar ice caps ('collapse') are based on a faulty understanding of the physics of ice flow, and also fail to account for the fact that major ice sheets are located in bedrock depressions. As Ollier and Pain have reminded us[209], the Antarctic ice sheet contains a record of ice accumulation almost a million years in length, and it is a physical impossibility for either it or the Greenland ice cap to collapse in the fashion asserted by the GCM modellers.

- Mountain valley glaciers are not uniformly retreating worldwide; instead, some are retreating and some are advancing[210]. In any case, from a climatic (sea-level change) point of view they are largely insignificant, containing only

Fig. 30 Twenty glacier length-records from different parts of the world, starting before 1600. Note that a general decline in glacier length commenced in the mid-nineteenth century, well before any possible influence from industrial carbon dioxide emissions, and that accelerated melting does not occur in the late twentieth century – indeed, some glaciers show a diminution of late 20th century retreat or even re-advance (Norway, New Zealand).

141

~180,000 km^3 (less than 0.6 per cent) of the world total of ~32 million km^3 of ice, for an imputed sea-level rise equivalent to just 45 cm should they all melt[211]. Valley glaciers are also very poorly known. Out of more than 100,000 glaciers worldwide[212], up to 2006 there had been only 3,480 annual mass balance measurements reported from 228 glaciers (heavily biased towards Europe). Though the World Glacier Monitoring Service reports continuing net glacial ice loss for the year 2007-08[213], that only applies to its small sample of 90 selected glaciers, and the time series only extends back to 1980. Overall, the available sparse data for the last 60 years 'indicate a strong ice loss as early as the 1940s and 1950s followed by a moderate mass loss until the end of the 1970s and a subsequent acceleration that has lasted until now[214].' In other words, ice retreat was underway before the great post-WWII increase in carbon dioxide emissions, and continued during the 1940-1975 phase of global cooling; it is therefore clearly not simply driven by either warming or by human emissions.

- In reality, the current phase of glacial retreat in the European Alps started in the 1860s, and thus long predated major industrial carbon dioxide emissions (Fig. 30). That retreat is now uncovering early Holocene tree stumps and human habitation artefacts, establishing that there is nothing historically unusual in the present disposition of those glaciers. Swiss researchers have also shown that glaciers in the European Alps were smaller during Roman times than they are now[215], and that 7,000 years ago glaciers disappeared almost completely during the Holocene climatic optimum[216].

- In other words, valley glaciers wax and wane on multi-decadal, centennial and millenial time-scales, and no evidence exists that their present, varied behaviour is unusual, falls outside long-term norms or is related to dangerous human-caused warming.

- The late twentieth century retreat of Mt. Kilimanjaro's summit ice-field was not a response to global or local warming, but resulted from a regional decline in precipitation, perhaps aided by warm-ground melting of snow and ice in the fumerolic Kibo summit crater[217].

- The number of polar bears is not decreasing; instead, in only two out of twenty delineated regions do polar bear numbers show a decline[218], the overall number of bears is estimated at an all-time high of more than 20,000[219], and the biggest threat to their health is the continued issuing of hunting licenses.

- The Great Barrier Reef (GBR) is not in any danger from 'global warming', for there is no trend of increased water temperature across reef waters over the last 27 years[220]. Instead, over the same time period the trend has been one of increasingly incessant scaremongering about the reef being killed by crown-of-thorns starfish, by overfishing, by soil runoff, by bleaching from global warming and most recently by trace amounts of herbicides. In actuality, GBR water quality has not changed measurably since pre-European times[221], and no scientific study yet shows significant overfishing or other human-caused damage to the GBR – outside of the very local, and strictly controlled, impact of tourist developments at a handful of reef sites out of more than 3,000 reefs[222].

- Were Captain James Cook to sail up the Queensland coast today out of sight of land, furnished with modern scientific measuring instruments, he would be unable to detect any changes in the state of the reef from when he first observed it in 1770.

Writer Michael Crichton certainly got it right when he said[223]:

> The greatest challenge facing mankind is the challenge of distinguishing reality from fantasy, truth from propaganda. ... We must daily decide whether the threats we face are real, whether the solutions we are offered will do any good, whether the problems we're told exist are in fact real problems, or nonproblems.

The null hypothesis

All of which brings us to the matter of Occam's Razor and the null hypothesis. William of Occam (1285-1347) was an English Franciscan monk and philosopher to whom is attributed the saying '*Pluralitas non est ponenda sine necessitate*', which translates as 'Plurality should not be posited without necessity.' This is a succinct statement of the principle of simplicity, or parsimony, that was first developed by Aristotle and which later came to underlie all scientific endeavour.

Following from this, the phrase 'Occam's Razor' is now often used as shorthand to represent the fundamental scientific assumption of simplicity. To explain any given set of observations of the natural world, scientific method proceeds by erecting, first, the simplest possible explanation (hypothesis) that can explain the known facts. This simple explanation, termed the null hypothesis, then becomes the assumed interpretation until additional facts emerge that require modification of the initial hypothesis, or perhaps even invalidate it altogether.

Given the great natural variability exhibited by climate records, and the failure to date to compartmentalize or identify a human signal within them, the proper null hypothesis – because it is the simplest consistent with the known facts – is that global climate changes are presumed to be natural unless and until specific evidence is forthcoming for human causation[224].

There are literally thousands of papers published in refereed journals that contain facts or writings consistent with this null hypothesis. Whether particular papers were written by 'alarmists' or 'sceptics' – who to defenders of the global warming faith appear to equate with angels and devils, respectively – is, of course, irrelevant. For science facts and interpretations are entirely independent of the good character or otherwise of those who describe or fund them, a characteristic that should be particularly noted by readers of postmodernist bent.

As explained in a late 2009 article by Martin Cohen[225], the world's very best advertising agencies have been employed to help the IPCC fashion the most effective climate alarmist messages possible. Such agencies are professionally both expert and thorough in devising ways to hoodwink the public[226], which they have indeed helped the IPCC to achieve. Thus counselled, one of the climate alarmists' most effective ploys has been to reverse the null hypothesis. In a perversion of scientific method, the onus of proof is then claimed to fall upon those who challenge the hypothesis of dangerous warming caused by human carbon dioxide emissions.

In reality, and because both the rate and the magnitude of recent warmings and coolings fall within the bounds of previous natural climate variation (Chapters 1, 2), the burden of proof of a human causation for change lies with those who would assert it rather than those who question it.

The single most important conclusion that can be drawn from the recent climate change for which we have accurate instrumental measurements, including expressly the mild late twentieth century warming, is that the null hypothesis that they have a natural origin remains unfalsified.

Tests of the dangerous warming hypothesis

No paper has yet been published that unambiguously invalidates the null hypothesis of a natural origin for observed, modern climate change, despite an estimated expenditure since 1990 of around US$100 billion (US$79 billion in the USA alone[227]) and the intense efforts of many scientists to find evidence that favours dangerous human-caused warming. These facts notwithstanding, it is an entirely legitimate exercise to propose and test alternative hypotheses also, in which regard the hypothesis favoured by the IPCC since the early 1990s has been 'that human greenhouse emissions (especially of carbon dioxide) will cause dangerous global warming.'

In 2007, the Australian Environment Foundation (AEF) invited me to speak to their annual conference in Melbourne on the topic of climate change. Choosing an analogy between the IPCC's alarmist stance and an ocean liner, my talk *Balance and Context in the Global Warming Debate* chose eight 'torpedos' as independent tests of the scientific and economic value of the dangerous warming hypothesis, all which the hypothesis fails. These tests are enumerated briefly in the end-note[228].

The global warming liner was thereby struck eight times, and just as it takes only one well-positioned torpedo to sink a ship so it should only take a single test failure to invalidate a scientific or policy hypothesis. Predictably, however, the Australian media ignored the 2007 AEF meeting, and the good ship 'Global Warming' sailed serenely on towards her eventual demise in Copenhagen in December 2009. Thanks to Victorian filmmaker Leon Ashby, however, a four-part version of the AEF lecture was posted on YouTube[228], where it subsequently drew wide attention – to date having attracted more than 300,000 viewings and being listed as the third most discussed Australian 'news and current affairs' item since YouTube coverage began ('science' apparently not being important enough to merit its own category). Clearly there is a public demand for independent assessments of the global warming issue.

This book is not the place to provide detailed discussion of the technical science issues, but it is nonetheless noteworthy that over recent years the great 'Global Warming' liner has endured yet more disabling scientific torpedo hits. Here is an abbreviated selection.

- Generating a potentially alarming degree of future warming using GCMs requires that there be an enhanced climate sensitivity for carbon dioxide increases, which the IPCC estimates to be 2.0-4.5°C for a doubling (4AR, SPM p. 885). Recent papers by Spencer and Bracewell (2008), Lindzen and Choi (2009) and Soon (2009) – using independent methods – confirm earlier independent estimates of a low sensitivity of 0.5°C or less for a doubling of carbon dioxide[229].

- The major warming effect projected by the IPCC comes not so much from increasing carbon dioxide per se, but from an assumed parallel increase in water vapour caused by enhanced evaporation. Actual measurements indicate precisely the opposite, with a decrease in pan evaporation measurements over recent decades[230].

- Ocean temperatures off, for example, southern Australia have been falling over the last several thousand years (Calvo *et al.*, 2009, from seabed core measurements)[231] but only recently, since the deployment of the Argo buoy network, have modern global ocean temperatures been measured with any precision. The surprising result has been the discovery that shallow ocean temperatures have been declining globally since about 2004[174].

- Several recent papers have demonstrated the dominant influence that the Southern Oscillation (El Niño/La Niña) exerts on mean global temperature. For example, McLean *et al.* (2009, 2010) have shown that, once the cooling perturbations caused by major volcanic eruptions are removed, global temperature faithfully tracks the Southern Oscillation index with a five to seven month delay, which leaves little room for a significant human forcing agent[232]. Instead 'overall the results suggest that the Southern Oscillation exercises a consistently dominant influence on mean global temperature, with a maximum effect in the tropics, except for periods when equatorial volcanism causes ad hoc cooling.' Furthermore, the stepped increase in global temperatures across the Great Pacific Climate Shift in 1976-77 resulted from the dominance after 1977 of warmer El Niño conditions over cooler La Niña conditions. Auckland climatologist Chris de Freitas, one of the authors of summary, commented that the research shows 'that internal global climate-system variability accounts for at least 80% of the observed global climate variation over the past half-century'. It may even be more if the period of influence of major volcanoes can be more clearly identified and the corresponding data excluded from the analyses[233].

- In their desire to attribute blame to carbon dioxide emissions, IPCC authors have consistently downplayed the effect that land-use changes have on local, and hence potentially global, temperature. Recent papers by McKitrick and Michaels (2004), Pielke *et al.* (2007), Compo and Sardeshmukh (2009) and Fall *et al.* (2009) demonstrate the importance of land-use change, and ocean-land heat balance, in determining temperature[46, 47, 234].

- Loaiciga (2006) has modelled the effect that doubling atmospheric carbon dioxide from 380 to 760 ppm would have on the pH of the ocean[235]. The results show a pH change of less than 0.2 units, which means that average ocean pH would remain within the currently defined water quality limits for seawater.

Other recently published research throws doubt on the IPCC's central claims. This includes[236] work that: challenges the validity of the concept of global temperature as a measure of climate change (Essex, 2007a; Essex *et al.*, 2007a); demonstrates the inadequacy of deterministic climate modelling (Essex, 2007b; Essex *et al.*, 2007b; Green *et al.*, 2009; Koutsoyiannis *et al.*, 2008; Kucharski *et al.*, 2009; Kukla & Gavin, 2004); highlights the importance of synchronized chaos (unforced internal climate shifts: Swanson & Tsonis, 2009; Stockwell & Cox, 2009); demonstrates that negative feedback may be caused by clouds (Cotton, 2009) or water vapour (Paltridge *et al.*, 2009), or to anti-persistence in lower tropospheric temperature increments (Karner, 2002); and confirms a short residence time of carbon dioxide in the atmosphere (Essenhigh, 2009). Many hundreds of other papers also contain either direct criticism of the dangerous warming hypothesis, or facts that are inconsistent with its truth[237], in direct contradiction of the silly, yet oft-repeated, mantra that 'there are no peer-reviewed scientific papers that disagree with the IPCC's dangerous warming hypothesis'.

Despite this cornucopia of scientific thinking that conflicts with the IPCC's computer-projected dangerous warming thesis, until 2009 public opinion remained strongly sympathetic to the IPCC and its recommendations. This indicates, as will be borne out in the next three chapters, that the global warming issue long ago ceased being a scientific problem. Rather, and as the IPCC and its supporters had always intended, since at least the turn of the twenty-first century global warming has been primarily a social and political issue.

Conclusion

Environmental changes that occur in response to local or global climate change are properly the subject of research investigation. But, as evidence for human-caused global climate change, such lines of reasoning on their own are entirely circumstantial. It can be estimated that Western industrialized nations currently spend around US$10 billion a year on climate change research and policy matters, with a cumulative spend since 1990 that must exceed US$100 billion. Despite the expenditure of this large sum, and great research efforts by IPCC-related and other (independent) scientists, to date no scientific study has established a certain link between changes in any significant environmental parameter and human-caused global warming[82, 112]. Conversely, and consistent with the null hypothesis, plausible natural explanations exist for all the post-1850 global climatic changes that have been described so far. Finally, abundant compelling scientific evidence exists that invalidates the IPCC's hypothesis of greenhouse warming.

Coda

'But do you believe in global warming?' the reporter still persists (meaning, of course, 'do you believe in dangerous global warming caused by human carbon dioxide emissions?').

'It depends,' I reply, 'for there are many different realities of climate change.'

In Chapters 7-10, we turn to the third and last of those realities, which is that of the social and political aspects of the current climate change debate.

7 Noble cause corruption

On the one hand, as scientists we are ethically bound to the scientific method, in effect promising to tell the truth, the whole truth, and nothing but – which means that we must include all doubts, the caveats, the ifs, ands and buts. On the other hand, we are not just scientists but human beings as well. And like most people we'd like to see the world a better place, which in this context translates into our working to reduce the risk of potentially disastrous climate change. To do that we need to get some broad based support, to capture the public's imagination. That, of course, means getting loads of media coverage. So we have to offer up scary scenarios, make simplified, dramatic statements, and make little mention of any doubts we might have. This 'double ethical bind' we frequently find ourselves in cannot be solved by any formula. Each of us has to decide what the right balance is between being effective and being honest. I hope that means being both.

(Stephen Schneider, *Discover Magazine*, 1988)[238]

No matter if the science is all phony, there are collateral environmental benefits… climate change (provides) the greatest chance to bring about justice and equality in the world.

(Christine Stewart, Environment Minister for Canada, 1998)[239]

It is part of the traffic of public discussion about global warming that some of the participants are corrupt. Routinely, climate scientists employed at even the most prestigious institutions are accused of having their alarmist views bought by a need to maintain research funding. Equally, self-righteous alarmist critics make desperate attempts to link scientists who express independent views with what are claimed to be the vested interests of the coal and oil and gas industries.

It is also obvious that commercial interests – including alternative energy providers such as wind turbine manufacturers, big utility companies such as Enron, financiers such as Lehmann Brothers, and the emerging group of emissions and carbon indulgences traders – have strong potential to become involved in corrupt dealings in the traditional meaning of the term. To varying degrees all of these accusations are true, but probably the strongest alarmist influence of all on the climate policy debate is the rather subtler phenomenon of noble cause corruption.

In his book *Science and Public Policy*[240], Australian Professor Aynsley Kellow explores the problem of noble cause corruption in public life in some depth. Such corruption arises from the belief of a vested interest, or powerful person or group, in the moral righteousness of their cause. For example, a police officer may apprehend a person committing a crime and, stuck with a lack of incriminating evidence, proceed to manufacture it. For many social mores, of which 'stopping global warming' and 'saving the Great Barrier Reef' are two iconic examples, it has become a common practice for evidence to be manipulated in dishonest ways, under the justification of helping to achieve a worthy end[221]. After all, who wouldn't want to help to save the Great Barrier Reef?

Regrettably, and perhaps driven in part by their support for broad environmental causes, not all scientists within the climate community have maintained the dispassionate, disinterested approach that is the *sine qua non* of good scientific research. Instead, some have become advocates for the cause rather than the science of global warming. This chapter discusses this, and some other examples of noble cause corruption.

The infamous 'hockey stick' curve of northern hemisphere temperature

Perhaps the most widely known piece of defective climate science of this type is the 1998 'hockey-stick' paper in *Nature* by Mann, Bradley and Hughes (MBH), and its 1999 and 2003 companion papers in *Geophysical Research Letters*[241].

The MBH papers were based upon statistical analysis of about 183 tree-ring records from sites across the northern hemisphere, based on the assumption that the width of individual rings bears a direct relationship to the temperature at the time of growth. The resulting graph of reconstructed temperature from 1200-1900, then projected by thermometer measurements to 2000, was dubbed the 'hockey-stick' because of its resemblance to a North American ice-hockey stick (Fig. 31). The graph exhibits a gently declining overall temperature from AD 1200 to 1900 (the handle) followed by a sharp rise in temperature thereafter (the blade).

The MBH curve played a central role in the launch of the IPCC's Third Assessment Report (TAR), appearing prominently as Fig. 1b in the Summary for Policymakers, Fig. 5 in the Technical Summary, and twice each in Chapter 2 (Figs. 2-20, 2-21) and the Synthesis Report (Figs. 2-3, 9-1B). In the Summary for Policymakers, the graph was accompanied by the comment 'that the 1990s has been the warmest decade and 1998 the warmest year of the millennium' for the northern hemisphere. This message was promulgated heavily at the release of the TAR in January 2001 and subsequently marketed heavily by environmental lobby groups. In short, the MBH hockey stick was seen to be of paramount importance for the IPCC case that dangerous human-caused climate change is occurring.

But all was not as it seemed, for the hockey stick curve contradicted earlier understanding in two main ways. First, it showed that temperatures held mainly steady, though with a slight long-term decline, between 1000 and 1900; i.e. known climatic episodes such as the Mediaeval Warm Period and multi-troughed Little Ice Age did not appear, despite the fact that many individual

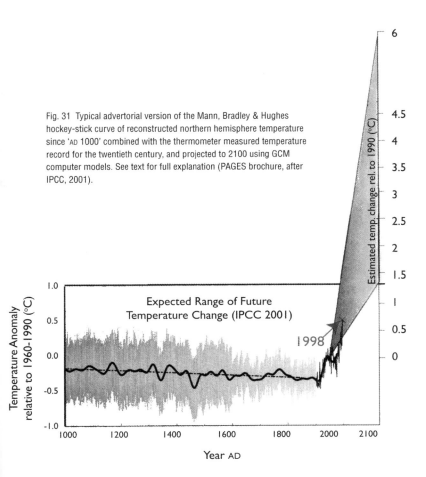

Fig. 31 Typical advertorial version of the Mann, Bradley & Hughes hockey-stick curve of reconstructed northern hemisphere temperature since 'AD 1000' combined with the thermometer measured temperature record for the twentieth century, and projected to 2100 using GCM computer models. See text for full explanation (PAGES brochure, after IPCC, 2001).

Mann et al., 1999; Houghton et al., 2001; Alverson et al., 2002

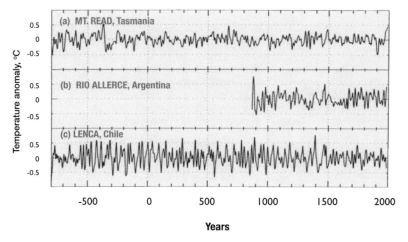

Fig. 32 Southern hemisphere tree ring records up to 2,800 years long from Tasmania, Argentina and Chile. Note the general variability about a long-term mean, and the absence of any unusual warming during the late twentieth century.

high resolution climate records depict these episodes with clarity (Fig. 6)[242]. Second, the rapidly ascending twentieth century blade of the curve appeared to show temperature increasing at an unprecedented rate and to an unprecedented magnitude by the end of the century, yet high quality speleothem and tree ring records from the southern hemisphere showed not a trace of the dramatic warming that was now being claimed for the northern hemisphere (Fig. 32). Nonetheless, the MBH graph gave birth to the legend that late twentieth century temperature was increasing to an unprecedented peak at a dangerous rate, the alarm being fanned even further by the appearance of versions of the hockey stick diagram with steeply rising computer GCM temperature predictions out to 2100 cobbled onto its top end.

Alarm bells immediately rang with those scientists who were familiar with statistics, for it is an error of the most elementary kind to combine in one graph actual physical measurements of temperature and inferred pre-historic temperature values calculated from measurements of proxy records like tree rings. In addition, many scientists noticed the conflict between the MBH hockey stick and other more direct climate records for the same period. After a brief pause, during which such independent scientists drew breath and thought about the matter, new studies of temperature change over the last 500-1500 years started being published that reaffirmed traditional interpretations, and conflicted sharply with the hockey stick curve. However, the initial charge against the hockey stick was led, and very much from left field, by the redoubtable Steve McIntyre, a Canadian mining analyst with a degree in mathematics and a forensic statistical mind[243].

McIntyre's experience with the analysis of speculative mining ventures – where rising hockey stick graphs of potential profits are a dime a dozen – led to a sharp twitching of his statistical nose. Accordingly, he combined with Dr Ross McKitrick, an economist from the nearby University of Toronto, to attempt to replicate the MBH hockey stick temperature graph. Because the authors and publishers of the original paper were grudging in their release of the data on which the research was based, the difficulty of trying to replicate their work lay not only in the advanced statistics involved. In time, though, enough data was crowbarred out into public view for the hockey-stick curve to be replicated and shown to be based upon incompetent statistical analysis (McIntyre & McKitrick, 2003, 2005)[244]. In particular:

- The MBH work contains basic data entry and statistical errors, including collation errors, unjustifiable truncation or extrapolation of source data, obsolete data, geographical location errors, incorrect calculation of principal components and other quality control defects.

- The earlier part of the MBH graph is dominated by the results from one data series only, that of bristlecone pines from western USA.
- When these errors were corrected, re-running the analysis yields a curve in which temperatures in the fifteenth century were higher than they are today.
- The principle components techniques used by MBH had a strong intrinsic tendency to produce 'hockey stick' shaped curves, even when working with random data.

Not surprisingly, a public furore arose around McIntyre and McKitrick's (MM) criticisms. That the MBH paper was flawed beyond retrieval was finally established by Professor Wegman, who undertook an independent audit of the work for a US Congressional committee that vindicated MM's analysis[245].

On being informed of the flaws in its 1998 MBH paper, *Nature* required a Corrigendum to be published. But they couldn't even get this right, allowing MBH to state, untruthfully, that 'None of these errors [that they were correcting] affect our previously published results' and refusing to publish an admonitory letter that identified this claim as untrue[246]. An extensive and detailed history of the entire MM interchange with MBH, which extended over several years, is provided here[247].

Whilst the MM versus MBH swordfight was playing out, other scientists were attacking the hockey stick climate history using other data sources and techniques. Studies that conflicted with the MBH interpretation of recent temperature history, and thereby adding to MM's demolition of it, included the following:

- Soon & Baliunas (2003)[248] showed that proxy data from throughout the northern hemisphere frequently contained climatic episodes that represented mediaeval warming (~ AD 1000) or later Little Ice Age cooling (~1600-1860). And many papers continue to accrue which document the existence of both these climatic episodes in widely distributed, far field sites including in the southern hemisphere[249].

- Von Storch (2004)[250] showed that the error bounds plotted by Mann *et al.* needed to be widened, because of the low accuracy of the proxy measurements that they had used; this correction has the effect of raising the upper bound of the 95 per cent error range across their diagram to a level well above the temperatures of the late twentieth century; effectively, the MBH methodology significantly underestimates the full range of temperature variability of the last 1000 years, and, within statistical error, it is simply not possible to assert that late twentieth century temperatures were warmer than those of earlier times.

- Esper *et al.* (2002)[251] published statistical analyses of other northern hemisphere tree ring records which showed that the late twentieth century temperature rise reached a peak at about the same level as temperatures during the Mediaeval Warm Period at ~ AD 1000.

- Independent techniques for reconstructing temperature over the last several hundred years, for example from borehole temperature records and deep seabed cores, produce results which conflict sharply with the MBH reconstruction, showing temperatures during the Mediaeval Warm Period to be warmer than today's and during the Little Ice Age to have averaged 0.2-0.7°C below present[252].

- A more recent and more accurate summary of average worldwide temperature for the last 2,000 years has been compiled from 18 high quality proxy climate records by US scientists Craig Loehle and Huston McCulloch (2008)[27], who demonstrate clearly the existence of the Mediaeval warming and Middle Ages cold episodes (Fig. 6), and capture also a rhythmic millennial periodicity that probably corresponds to the Bond solar cycle (Chapter 1).

Attitudes to data release
One of the problems that confronted MM in their attempts to replicate the hockey stick graph was that the MBH authors

obstructed the release of the full dataset and computer codes on which their 1998 paper was based. Amazingly, neither the research agencies that funded the research nor the journals that published it were prepared to insist upon the fundamental principle that applies to all publicly-funded research, which is that an author's data and a full description of the techniques used must be made readily and transparently available to other scientists who wish to attempt replication of the results.

This is not a new problem, since the same issue had surfaced during the preparation of the IPCC's Second Assessment Report (2AR), when a reviewer of part of the draft requested that he be supplied with some of the raw data on which the work was based. The author, Dr Tom Wigley, declined to supply the data, making the following astonishing statement[253]:

> First, it is entirely unnecessary to have original 'raw' data in order to review a scientific document. I know of no case at all in which such data were required by or provided to a referee ….. Second, while the data in question (model output from the UK Hadley Centre's climate model) were generated using taxpayer money, this was UK taxpayer money. US scientists therefore have no *a priori* right to such data. Furthermore, these data belong to individual scientists who produced them, not to the IPCC, and it is up to those scientists to decide whom they give their data to.

This reply denies the supply of data to another scientist who wishes to check that the work can be replicated; denies data to a scientist on the grounds that he is from another country; and arrogates to the author the right to decide who, if anyone, would be supplied with data which was collected with public funds and which underpins an important international publication. Each one of these actions constitutes a fundamental breach of science etiquette, and were such attitudes to be promulgated widely, science as a value-free, objective, internationally agreed enterprise would collapse.

That, nonetheless, such attitudes are widespread within the IPCC climate science community was revealed in devastating fashion in November 2009, with the leaking of a large file of emails and related documents written by research staff at the UK's Climatic Research Unit (CRU), as described in more detail in Chapter 12.

Government agencies and reports

Regrettably, it is not just individual scientists who are involved in trying to control the climate change debate by the use of selective science. Scientists who work for major governmental science agencies in Western countries are almost all under strict employer instruction as to public comments that they may make about climate change, always remembering that a substantial slice of their budget is provided for global warming research.

For example, Australian science journalist Peter Pockley reported in 2004 that the

> Commonwealth Scientific and Industrial Research Organization's [CSIRO] marine scientists have been 'constrained' on the scientific advice and interpretation they can provide to the government's conservation plans for Australia's oceans. Likewise, climate scientists have been told not to engage in [public] debate on climate change and never to mention the Kyoto Accord on greenhouse gas emissions.

A more recent example was CSIRO CEO Megan Clark's extraordinary choice of language when announcing that the Australian science organization would 'punish' economist Dr Clive Spash for delivering without permission a conference paper that criticized the government's intended emissions trading scheme[254]. That the paper had in the meantime been accepted for publication in the peer-reviewed international journal

New Political Economy, and also been tabled in parliament, was no protection for the hapless Dr Spash who was eventually forced to resign.

In Australia, CSIRO is the government agency that makes most of the public running in the global warming debate, and that almost exclusively on the side of environmental alarmism. CSIRO's GCM modelling group, which acts as a science provider to the IPCC, exploits its resulting near-monopoly situation in Australia by providing also climate change consultancy reports to state and federal governments, regional authorities and planning boards and large industrial organizations. In turn, these organizations develop restrictive public regulations based on the false presumption that they have been provided with accurate information about how climate will change in the future. The CSIRO reports, of course, are based around regional GCM modelling that produces unvalidated projections of future climate, not predictions or forecasts. With astonishing panache, they contain the following disclaimer:

> This report relates to climate change scenarios based on computer modelling. Models involve simplifications of the real processes that are not fully understood. Accordingly, no responsibility will be accepted by CSIRO or the QLD government for the accuracy of forecasts or predictions inferred from this report or for any person's interpretations, deductions, conclusions or actions in reliance on this report[255].

The subtlety of a fine distinction between a climate projection and a climate prediction escapes CSIRO's clients. Together with CSIRO staff, they make no attempt to correct or check the media's unvarying presentation of such model results as if they were firm predictions, and their subsequent association with rampant public climate alarmism.

In these several ways, science policy advice coming from

government departments and agencies is routinely corrupted to suit the views of the government of the day. In turn, governments's views are often strongly influenced by noble cause corruption, whereby 'saving the planet' is seen cynically as an effective way in which to harvest green and swinging votes, quite irrespective of the lack of demonstrated, as opposed to interest group promulgated, risk. The inaccurate and alarmist global warming television advertisements that in 2009 both the British and Australian governments used taxpayer funds to run are very much cases in point.

Governments have come to their views about global warming in response to electorates that in turn are responding to a long and well orchestrated propaganda campaign by environmental organizations. In a viciously self-reinforcing cycle, this has led to excessive government spending on climate-related research (with all its corrupting influence) and the widespread employment of public relations professionals and advertising agencies by science organizations, governments and environmental NGOs alike – all of whom, together with the press, have a vested interest in perpetuating the warming scare in what Richard Lindzen has termed the iron triangle of alarmism. The process works like this: (i) scientists make meaningless ('ten out of the last eleven years are the warmest on the meteorological record') or ambiguous ('global warming is a fact') statements; (ii) advocates and the media translate these statements into alarmist declarations that are often awash with *non sequiturs* or deliberate obfuscation ('global warming is human-caused and demands reductions in carbon dioxide emissions'); and (iii) politicians respond to the alarm, and harvest votes, by feeding scientists more money. As Richard Lindzen notes[256]:

> But there is a more sinister side to this feeding frenzy. Scientists who dissent from the alarmism have seen their grant funds disappear, their work derided, and themselves libelled as industry stooges, scientific hacks or worse. Consequently, lies about climate change gain credence

even when they fly in the face of the science that supposedly is their basis.

Powerful, overarching political spin can be generated once an obedient covey of scientists with their 'rice begging bowls' has been captured within Lindzen's iron triangle. That spin is then disbursed into the public realm through conferences, through the education system and through the media, as discussed in the next chapter.

The 'sinister' part of this is the intimidation of scientists involved, which is immensely strong. In the UK, for example, the BBC, commercial television, all major newspapers, the Royal Society, the Chief Scientist, the Archbishop of Canterbury, the Bishop of London, David Attenborough, countless haloed-image organizations such as the Royal Society for the Protection of Birds, and even Prince Charles himself all stridently assert an alarmist line on global warming, in the face of which it is simply professional suicide for a scientist to put a questioning head above the parapet. Though a similar scenario applies in other Western countries, the British trait of deference to 'old' authority makes the situation particularly bad there.

Political climate reviews: the Stern and Garnaut Reports

Public discussion on climate change in the UK since 2006, and in Australia since 2008, has drawn heavily on climate reviews prepared for government by Nicholas Stern and Ross Garnaut, respectively, and these reports have achieved worldwide resonance. The two economists faced the task of advising their governments on the issue of human-caused global climate change in the context of the presence, or likely introduction, of carbon dioxide taxation (aka emissions trading schemes). After the publication of Garnaut's report, the new Australian Labor government came up with the positively Orwellian title of the *Carbon Pollution Reduction Scheme* for their proposed new system of carbon dioxide imposts. In a misbegotten attempt to explain this scheme, minister of climate,

Senator Penny Wong, achieved what must be a world record of encapsulating no fewer than seven scientific errors in the first sentence of her Green Paper on the matter[257].

Obviously, both Stern and Garnaut faced making an initial judgement as to whether late twentieth century warming was continuing (as asserted by some scientists) or whether cooling had set in (as asserted by others) and, in either case, what were the likely causes. These being scientific judgement calls, to appoint economists to adjudicate upon them puts the reviewers in the invidious position of having to base their economic recommendations upon science authority provided by others. The convenient authority at hand was, of course, the IPCC, whose politically-tainted 'science' advice both Stern and Garnaut swallowed whole with nary a blink. The science in both of their reports is simply a subset of that in the IPCC Fourth Assessment Report, in some cases spiced up by selective use of other references[258].

The policy prescriptions outlined in both reports was therefore predicated on two assumptions. First, that global temperature is rising (implicitly, at either an unusual rate or to an unusual magnitude); and, second, that adding more carbon dioxide to the atmosphere will result in dangerous warming. Both of these assumptions are self-evidently wrong – notwithstanding that global temperature does from time to time warm (and cool), and that carbon dioxide is a minor greenhouse gas.

That Garnaut and Stern's economic analyses were not only erected upon a faulty science edifice but also contain major flaws in their economic reasoning was pointed out shortly after their publication[259], causing former UK chancellor Nigel Lawson to peremptorily dismiss Stern's report as 'quite the shoddiest pseudo-scientific and pseudoeconomic document any British Government has ever produced'[260], a judgement that is supported by recent revelations that covert alterations were made to the report between its initial release in October 2006 and its formal publication in January 2007 by Cambridge University Press[261]. Among the alarmist claims removed, or errors corrected, were those relating to the occurrence of more cyclones in northwest Australia, the reduction of rainfall in southern Australia and an order of magnitude error made regarding the costing of US hurricane damage. Predictably, though the unfounded Australian alarms were widely reported in the press at the time of their October release, there was no subsequent correction. Therefore, and despite their political popularity, the recommendations of these deeply flawed reports have little relevance to real world policy formulation. The problem is exacerbated because both reports also ignored the pressing issue of hazardous natural climate change and the contextual fact that global temperature is now cooling and is predicted, by many scientists, to continue to cool. Above all else, planning for future climate hazard has to be based on a thorough, realistic risk analysis, and the Stern and Garnaut reports utterly failed to provide this.

An accomplished cost-benefit analysis of climate change requires two things: a clear, quantitative understanding of the natural climate system and a dispassionate, accurate consideration of all the costs and benefits of cooling as well as warming. Unfortunately, neither the Stern or Garnaut reviews provide this. They are not cost-benefit analyses but rather contrived risk analyses of warming only, and adroitly shuffle the peas under the thimble to provide flawed and partial accounts of our possible climatic future that stress the costs yet largely ignore the benefits of warming, and fail to consider properly the likely eventuality of future cooling.

Though they will doubtless continue to be lionised for some time yet, while it remains politically expedient to do so, the Stern and Garnaut reviews are destined to join Paul Ehrlich's *The Population Bomb* and think tank The Club of Rome's manifesto, *Limits to Growth*, in the pantheon of big banana scares that proved to be without foundation. They are part of the last hurrah for those who inhabit the world of virtual climate reality that exists only inside flawed computer models.

Science academies and learned societies

Traditionally, governments wishing for dispassionate advice on a science issue have turned to their nation's science academy. Disturbingly, against this historic context, *Nature* formerly reported that in appointing one of its former presidents, a high-profile former government adviser (Lord Robert May), the Royal Society was intensifying its moves into the public and political arena – and taking a calculated risk. 'If you want to be more effective in engaging issues of public concern, then you really need to understand the rules of engagement,' said May[262].

The path being followed duly became evident when in 2001 Lord May helped organize a statement published in *Science* magazine that there was a scientific consensus on the danger of human-caused global warming. The statement was headed *17 National Academies Endorse Kyoto*, and May was reported as commenting that it had been partly provoked by (President George) Bush's recent rejection of the Kyoto treaty, along with resistance to the Kyoto terms by countries such as Australia[263].

Many scientists were astonished to find the historic Royal Society playing such naked political games. Added to which, for a scientific academy to argue that scientific truths, which are discovered by hypothesis testing, should instead be established by the political process of consensus, is a betrayal of all that science stands for, and ridiculous to boot.

Against this unhappy background, it shouldn't be surprising, but is, to discover that in 2006 the Royal Society's policy communication officer, Bob Ward, wrote an intimidatory letter[264] to oil company Esso (UK) in an effort to suppress Esso's funding for organizations that in the Royal Society's view 'misrepresented the science of climate change, by outright denial of the evidence that greenhouse gases are driving climate change … or by overstating the amount and significance of uncertainty in knowledge, or by conveying a misleading impression of the potential impacts of anthropogenic climate change.' Ward's attempt to prevent free public discussion of global warming resulted in rapid condemnation, including a comment from the US Marshall Institute[265] that:

> It is indeed unfortunate that the Royal Society is advocating censorship on a subject that calls for debate. The censorship of voices that challenge and provoke is antithetical to liberty and contrary to the traditions and values of free societies. That such a call comes from such a venerable scientific society is disturbing and should raise concerns worldwide about the intentions of those seeking to silence honest debate and discussion of our most challenging environmental issue – climate change.

Notwithstanding widespread condemnation of the Royal Society action, copycat attempts to intimidate other businesses soon followed. In the USA, Senators Rockefeller and Snowe wrote an intimidating letter to Esso's partner company Exxon; and in Australia, a Labor shadow minister, Kelvin Thomson, basing his views on a showing of Al Gore's film and the Royal Society letter, wrote in similar fashion to a number of leading Australian companies.

The Royal Society continued its political pursuits through to 2005, when the 31st G8 meeting was hosted by UK prime minister Tony Blair at Gleneagles, Scotland. Like pulling a rabbit out of a hat, and with exquisite timing about one week before the conference, the Society released to the press an alarmist statement on human-caused global warming, but this time with the signatures of 12 other national science academies attached. Within days, both the US Academy of Science and the Russian Academy of Sciences had protested at the inclusion of their signatures. Undeterred, the Royal Society and its G8+5 group of partner national science organizations in Brazil, Canada, China, France, Germany, Italy, India, Japan, Mexico, Russia, South Africa and USA marched militantly on towards similar alarmist press releases in 2007 and

2008 (before the 33rd and 34th G8 summits) and most recently in 2009 (before the COP-15 Copenhagen meeting).

Meanwhile, the disease of sanctimonious public proselytising proved contagious, and since about 2005 the councils of many, many professional groups and societies all over the world have rushed to add their particular society's name to what they perceived to be a worthy and illustrious list[266]. Astonishingly, in view of the hotly disputed nature of much of the science involved, more than 50 such professional organizations have now released statements proclaiming their alarm about global warming, and urging governments to take action to curb it.

Nor are things any better across the Atlantic, where Lindzen has described in chilling detail the way that the US National Academy of Sciences has been infiltrated by environmental activist scientists. As Lindzen recounts it[267]:

> The Academy is divided into many disciplinary sections whose primary task is the nomination of candidates for membership in the Academy. Typically, support by more than 85 per cent of the membership of any section is needed for nomination. However, once a candidate is elected, the candidate is free to affiliate with any section. The vetting procedure is generally rigorous, but for over twenty years, there was a Temporary Nominating Group for the Global Environment to provide a back door for the election of candidates who were environmental activists, bypassing the conventional vetting procedure. Members, so elected, proceeded to join existing sections where they hold a veto power over the election of any scientists unsympathetic to their position.
>
> Moreover, they are almost immediately appointed to positions on the executive council, and other influential bodies within the Academy. One of the members elected via the Temporary Nominating Group, Ralph Cicerone, is now president of the National Academy. Prior to that, he

was on the nominating committee for the presidency. It should be added that there is generally only a single candidate for president. Others elected to the NAS via this route include Paul Ehrlich, James Hansen, Stephen Schneider, John Holdren and Susan Solomon.

The concluding names here are a pedigree list of climate alarmists if ever there was one, and once such persons became influential inside the NAS the organization's official views on global warming became closely aligned with those of the Royal Society of London.

What none of these science organizations tell the public, of course, is that the alarmist statements that they release are carefully contrived for political reasons by their management and governing bodies. Damningly, so far as I am aware not one instance exists where a society's governing body has consulted with its expert membership prior to releasing a statement of alarm about global warming; and there are many examples of memberships later expressing extreme discontent about, or direct opposition to, alarmist statements that have been claimed to represent their views[268]. Beware the managers of scientific societies, like all others, for they like to manage; and beware even more the directors, for they like to direct; and neither of these are scientific activities.

The malignant influence of environmental organizations

The wider public has been gulled into believing that dangerous global warming is likely to occur by scientific propaganda and speculative computer models, in place of being informed with accurate factual information about real climate change. In addition to the sad role played in this by some members of the professional scientific community, as outlined earlier in this chapter, there are two other powerful causes of climate misinformation. They are the lobbying activities of environmental nongovernmental organizations (NGOs) and their allies, and the increasing inadequacy of education systems to impart an objective science education (see Chapter 8) – two topics that are actually intertwined.

Nearly all readers will be aware of the activities of high profile environmental NGOs such as Greenpeace, the World Wide Fund for Nature, the Sierra Club and the Australian Conservation Foundation. But relatively few persons appreciate the size, scope, co-ordination and colossal financial resources that are now involved in environmental lobbying around the world.

It is difficult to find accurate figures regarding the amount of money available to the large environmental NGOs, but indicative amounts can be estimated. For example, it was reported in the press that in 2002 the US branch of WWF alone turned over US$485 million, and the US branch of Greenpeace another US$242 million. Remembering that these organizations have other large branches throughout the world, and the existence of literally thousands of other similarly minded environmental groups, an estimate of the summed 2002 annual income for this industry of around US$2 billion seems not unreasonable. In that same year, Australia – a mid-sized wealthy, Western nation with a current population of around 20 million and a GDP of around $1 trillion – had an election campaign in which the three major political parties spent AUS$60 million on advertising. Attributing 10 per cent of that spending to environmental matters (which will be an overestimate), and noting that Australian elections occur once every three years, Australia was spending about AUS$2 million/year on environmental advertising.

Thus in 2002 the major environmental NGOs, who are focused on creating alarm about human-caused global warming, outspent the government of a medium-sized, wealthy Western economy by at least 1000 times. Is it any wonder that there is a public perception that human-caused global warming is a problem?

Driven by their addiction to alarmism, and with a false belief that the causes of climate change are understood, environmentalists worldwide urge the adoption of the precautionary principle to solve the 'global warming problem'. The reality that you can't take precautions against a future that is unknown (i.e. may encompass either warming or cooling, or both) is ignored in favour of irrational feel-goodery, the aim being to move the world to a 'postcarbon' economy by drastic curtailment of the carbon dioxide emissions that are alleged to be causing warming. And lobbying powerfully behind the scenes at every climate policy debate is to be found the little publicized Climate Action Network (CAN); founded in 1989, this powerful umbrella organization has seven regional nodes scattered across the globe, and co-ordinates the advocacy of more than 280 separate environmental NGOs[269]. CAN's vision is stated to be 'to protect the atmosphere while allowing for sustainable and equitable development worldwide'.

Conclusion

Across the world, many scientific and science administration organizations have caught the disease of noble cause corruption, and have become politicised over global warming and other contentious environmental issues. The organizations involved – or at least their governing bodies – have lost their way in a noxious (to science) fog of political correctness. Regrettably, governments and societies have now lost what used to be an important conduit of impartial and independent advice on technical matters of the day.

Those who campaign for reducing carbon dioxide emissions often believe that 'the ends justify the means' and 'whatever it takes' applies. In the absence of scientific evidence for dangerous human-caused warming, they cling blindly to the precautionary principle (Chapter 10). Normal moral principles are suspended, and both rigorous risk analysis and commonsense are ignored.

The campaigners remain blind to the inconvenient facts that, irrespective of the nobility of their cause: no amount of precaution is going stop natural climate change; there is a 100 per cent risk of damage from natural climate events, which happen every day; we cannot measure, much less isolate, any presumed human climate signal globally; extra atmospheric carbon dioxide causes mild warming at best, and, being highly beneficial for plant growth, helps to both green the planet and feed the world; and that the causes of climate change are many, various and very incompletely understood.

8   Communicating the story

> The task of climate change agencies is not to persuade by rational argument.... Instead, we need to work in a more shrewd and contemporary way, using subtle techniques of engagement.... The 'facts' need to be treated as being so taken-for-granted that they need not be spoken.
>
> Ultimately, positive climate behaviours need to be approached in the same way as marketeers approach acts of buying and consuming.... It amounts to treating climate-friendly activity as a brand that can be sold. This is, we believe, the route to mass behaviour change.
>
> (G. Ereaut & N. Segnit, 2006)[270]

The report from which the startling quotation above is derived, written by professional sociologists at the Institute for Public Policy Research in London, outlines the techniques that were employed by former UK prime minister Tony Blair in his relentless crusade against global warming. 'Amen to that', Al Gore would presumably sing.

Not chilled by statements like these? Then your global warming fever is likely incurable, for rarely has the public prostitution of an important science issue been so clearly revealed as in these inadvertent slips of the post-modernist skirt. Yet obnoxious spin mongering of this type is now a pervasive part of the way that most major Western political parties operate (and not just with regard to the global warming issue). The needed message is conveyed through the conduits of conferences, the education system and the media, and we will consider each in turn.

Climate conferences

The IPCC merry-go-round
A complex series of international workshops and meetings underlies the preparation of an IPCC assessment report. Since completion of the Fourth Assessment Report (4AR) in 2006- 2007, benchmark IPCC conferences have been held in Nairobi (2006), Bali (2007), Poznan (2008) and Copenhagen (2009). Meanwhile, between these dates a large number of other non-IPCC meetings have been held in many different countries, generally under the auspices of distinguished societies or professional organizations. Such meetings are invariably well funded; are often held in exciting or exotic locations; usually involve the support in various ways of the government of the country in which they are held; often involve the support (one might even say, heavy influence) of various European diplomatic commissions, especially the UK; invariably involve senior IPCC or IPCC-connected scientists; and usually find ways to exclude or minimize the alternative viewpoints about climate change that are offered by independent scientists.

I provide a brief description below of one such meeting, the *Climate Change and Governance Conference*, that was held in Wellington, New Zealand in March 2006. It was an all-too-typical example of the genre.

The 2006 New Zealand climate conference
Between 1999 and a general election held in December 2005, New Zealand was governed by two Labour-led coalition governments. The second of these coalitions (2002-2005) gave strong endorsement to New Zealand's membership of the Kyoto Accord. After an election in late 2005, a third Labour coalition was formed but without membership of the Green Party, a former partner. The new government announced that it would not be proceeding with a long-planned carbon tax that had related to New Zealand's Kyoto obligations. This presented the Green Party with the opportunity to argue that 'there is a gaping hole in climate policy, and it must

be filled.' The government moved towards formulating a new climate policy in three ways.

First, in mid-2005 the previous government had already initiated an inter-departmental review on climate change, which reported in November just prior to the election. This review was concerned primarily with identifying policies through which New Zealand could meet its obligations under two United Nations agreements: the Framework Convention on Climate Change (ratified by New Zealand in 1993) and the Kyoto Protocol (ratified in 2002). No substantial consideration was given to the strongly disputed science of climate change. Rather, the review team accepted *a priori* the adjuration of environment minister Pete Hodgson that 'the science is settled', and therefore dealt only with policy formulation.

Second, building on the inter-departmental report, early in 2006 the Ministry for the Environment held a series of two hour-long 'public engagement' meetings on climate change in New Zealand's three largest cities. The meetings were intended 'to provide information on the current status of Climate Change policy in New Zealand and to give people an opportunity to express their viewpoints.' Presenters included the chief executive officer of the ministry, and three other senior officers.

The Ministry's presentation started with the bullet point 'Everybody's climate is changing and will keep changing', a vacuous statement which is both entirely true and ever has been. The statement was supported by a slide that displayed an increasing global average temperature from 1900 to just past 2000, then projected on upwards to achieve a 2-3°C warming by 2100 according to two different growth scenarios. These computer temperature projections, of course, are entirely speculative and have no proven predictive skill. Yet ministry staff not only used this diagram as the background for their discussion, but they also presented a list of specific climate changes that were 'predicted' for New Zealand by 2080 by similarly unskilled computer models. (As senior National Institute of Water and Atmosphere, NIWA, climate scientist Jim Renwick has remarked about such models[271]: 'Climate prediction is hard, half of the variability in the climate system is not predictable, so we don't expect to do terrifically well.' Dr Renwick was responding to an audit showing that the long-term climate forecasts issued by NIWA were accurate only 48 per cent of the time; in other words, one can do just as well by flipping a coin).

Overall, these 'public engagement' meetings were clearly intended to lead public opinion to the conclusion that anthropogenic climate change is an established and dangerous hazard. The discussions took place in a virtual reality climate universe where real-world, scientific assessment was ignored, and – as two participants put it – the overwhelming requirement to be met was that New Zealand be seen to be 'a co-operative global society' and to 'portray the stance of global responsibility.'

As a third step towards developing a new climate policy, the government then gave sponsorship to the Wellington *Climate Change and Governance* conference, which followed closely, and probably not coincidentally, in the footsteps of the inter-departmental report and the public engagement meetings. The conference was sponsored by several New Zealand government departments (including environment), foreign embassies[272], the Royal Society of New Zealand, Victoria University and a range of greenhouse and business interests

As the organizer of the meeting, Professor Jonathon Boston, noted, 'This is a policy conference, not a science congress or a diplomatic negotiation.' The conference website upped the ante with the unsubstantiated assertion that: 'There is now little doubt that climate change represents one of the greatest and most urgent challenges faced by the world community,', and noted that the focus of the conference was to be 'on framing the governance and policy issues, and the scope for global, regional, national and local initiatives to reduce and manage the risks inherent in climatic shifts.' Any remaining doubt that the organizers viewed the science as 'already settled' was removed by the briefing paper distributed to participants before the conference. This stated that 'the policy debate can start from the general proposition that human activity, particularly in the post-industrial era, has changed the composition of the planet's atmosphere to the point where more serious consequences have become probable, and possibly inevitable.'

The list of speakers for the conference included many scientists from government or international science agencies that are identified with the alarmist global warming cause, together with senior environmental bureaucrats and activists. The high profile scientists and activists who attended received much media coverage, and some also gave talks outside the conference venue, including in Christchurch and Auckland. They included Dr Kevin Trenberth, head of climate analysis at the US National Center for Atmospheric Research, Professor David Vaughan, from the British Antarctic Survey, Lord Ron Oxburgh, a geologist and former chairman of Shell UK, and Kirsty Hamilton, formerly of Greenpeace. Despite the presence in New Zealand of at least four well-qualified and publicly active climate rationalists, who are known internationally for their constructive critical analyses of the complex issues of climate change, not one was invited to participate in the Wellington conference.

It is self-evident that the organizers of the Climate Change and Governance conference proceeded from a preconceived belief that dangerous, human-caused climate change is already proven. Their interest in learning about the science of climate change proved to be restricted to only that science which provided succour for their belief. One person who attended provided the following perception of the conference:

> Though individual speakers gave caveats about the science uncertainties that became totally irrelevant to the take-home message that emerged. Which was that climate change is here, is caused by human greenhouse gas emissions, and is changing faster than anyone had

previously thought it would. So the need for action is urgent.

The 2006 Wellington global warming conference is described in more detail here[273]. Dismayingly, it is typical of many other, similar conferences, both past and present. Together with the preceding inter-departmental review and public engagement meetings, it provides a textbook study in how modern governments seek to use selective science results from carefully selected advisers to indoctrinate the public with a viewpoint that meets the political imperative of the day.

Specialist professional meetings

In addition to politically contrived conferences like the 2006 Wellington meeting, many other more specialist workshops and meetings on specific aspects of climate science have also come to fall under the influence of global warming alarmism. Sometimes this influence is subtle, and sometimes it is not, but most scientific societies now have global warming activists as influential members whose aim is to slant the outcome of any climate-related meeting towards support for the dangerous, human-caused warming hypothesis.

A typical recent case has been documented by Dr Willie Soon and Dr David Legates[274]. In mid-2009, the organizers of the annual Fall 2009 Meeting of the American Geophysical Union (AGU) in San Francisco accepted a proposal from these two scientists to co-chair a session on the topic *Diverse Views from Galileo's Window: Researching Factors and Process of Climate Change in the Age of Anthropogenic CO_2*. A little later, and after arrangements had been approved to include within the Galileo session another 12 papers from a separately planned session on *Solar Variability and Its Effect on Climate Change*, chaired by Dr Nicola Scafetta, the AGU Planning Committee abruptly cancelled the *Galileo* session and moved its papers into other placements on the program.

To a non-scientist, this will sound like a storm-in-a-teacup squabble, but it is not. Rather, the matter represents a high-handed and successful attempt to subvert the original, and entirely sensible, scientific intention of Soon and Legates – which was to run a session that compared solar variability and human-related carbon dioxide emissions as factors in contemporary climate change. Truly independent scientific assessments of the global warming hypothesis, such as this proposed *Galileo* session at AGU, have been dying the death of a thousand such cuts over the last ten years. As Soon and Legates wrote at the time: 'The AGU should be ashamed. Its members should be outraged'; sadly, neither appears to be the case.

The education system

An increasingly powerful source of misinformation on climate change since the early 1990s has been the education system, at all levels. To what extent government, through regulation, education or any other means, should attempt to influence an individual's personal behaviour is a contestable issue. But one might have hoped that even those who support attempts to change adult behaviour on matters like smoking or obesity would object to the indoctrination of school children on matters of science. That, nonetheless, such indoctrination is now rife is not accidental, but stems from the relentless implementation of the tactics implied by the old Jesuit motto '*Give me the child until he is seven, and I will give you the man*'.

The process is illustrated by the following statement, from an educational consultancy in Australia[275] that believes that climate change is an education 'hot spot' and needs a nationally co-ordinated approach to its teaching:

> Climate change is here with us now and educational programs that are aimed at changing behaviour in relation to resource use with consistent national goals and clear messaging can provide an important and effective link to millions of homes in Australia.

Nanny clearly knows best, and the breadth of this problem is further illustrated by the five following letters. All were sent to me in response to articles and lectures in which I have questioned the prevailing mantra on human-caused climate change, and attempted to introduce some balanced commonsense into the public discussion.

Letter 1 – From a School pupil's parent

I appreciated your recent article regarding the global warming hysteria. My children have been filled with dread by all of the hype coming through our school system, reinforced by interminable media support. I have patiently explained things as best I can, but being a simple grocer from the Midwest (USA) I have less credibility than the experts.

I'd appreciate a bit of steerage towards some good information (in addition to your article) that will help me refute the current pop-culture pseudo-science.

Letter 2 – From a University student

My classes in Geology taught me how warming AND cooling trends occur throughout the history of the planet. Why is it so hard for people to understand that the planet is old and has had many episodes of climatic change?

I write to you today, because I am now in my junior year at a technical college in Virginia – I am enrolled in several Urban Affairs and Planning courses – one in particular is Environmental Planning. Time and again, I find myself at odds with nearly every one of my 55 classmates, in addition to the instructor, over the issue of global warming. Your article gives another voice to my own opinions. I guess my being 40+ years old in the mix of 20

year olds that surround me, perhaps may have something
to do with it as well.

Letter 3 – From a civil servant at the start of a professional career

Earlier this year, the *Financial Times Magazine* contained
an article about Sir Gus O'Donnell, Secretary to the
Cabinet, Head of the Civil Service and former Permanent
Secretary to H. M. Treasury. Here is the rhetorical
question that he is reported as posing to an audience of
newly-joined civil servants (Whitehall's newest and
brightest recruits):

'When you go to dinner parties do you want to be able
to say that you work with an accountancy firm and you've
spent the day helping some company pay less tax? Or do
you want to say you've been working with the
environment department to help save the planet, or with
the G8 group of countries to reduce child poverty?'

Letter 4 – From a Generation-Y journalist

I was at the conference you spoke to last Saturday. I was
most interested in your talk, mainly because I'm a Gen-Y'er
who has never heard an argument mounted against
'climate change'.

I should clarify – when I say I've never heard the other
side, I probably mean that I've never heard it through the
mainstream media. I'm a journalist and among my peers
almost everything is questionable these days. But I must
confess that it's social suicide to question climate change
– more than some other taboo topics, even abortion.
People can respect you for all manner of views, but no
one has time for a climate change 'denialist', as the label
goes.

Letter 5 – From a mid-career university climate teacher and researcher

> The reason for the apparent absence of public support from scientists for a sceptical position on climate change is professional intimidation.
>
> I have experienced it: letters to my vice-chancellor saying that I am irresponsible, snarly comments from students, personal threats from green wackos, condemnation by a national newspaper columnist (which caused me to take legal action), scowls and derision from all the folk who work for the government's main advisory department on climate change, and contempt from those of my colleagues who depend upon the research money on which global warming hysteria feeds.

These examples were sent to me from four different countries; thus the indoctrination and intimidation problems that they exemplify are global in scope. Those scientists and school and university teachers who wish to avoid unemployment will clearly not want to appear to take a balanced position on the climate issue, not publicly anyway. That doubtless many teachers quietly, and out of public gaze, continue to strive to provide accurate information to the young persons in their care does not remove the reality of the strong, and in many cases successful, pressure on all teachers to sing the siren hymn of climate alarmism.

Climate change is now taught in school using the multitude of glossy publications and websites that are provided by alarmist interest groups, which include many government departments. Choosing three examples at random, all aimed at primary school children, any reader who doubts the argument that I am relating here should take a trip to the following websites, one each supported by the Australian Broadcasting Corporation (ABC), the New Zealand government and a major UK power company, NPower[276].

The example from the ABC, Australia's public broadcaster, provides a 'greenhouse gas calculator' where, by answering questions about their lifestyle, a primary school child is encouraged to work out the amount of carbon dioxide that their activities produce, and thereby – and I am not making this up – 'to find out what age that you should die at so you don't use more than your fair share of Earth's resources.' The site has attracted strong protests from around the world, among them the apt comment that it represents child abuse. Our second example, from the New Zealand government, uses clever cartoon imagery and prejudiced text (for example, 'Our factories use a lot of energy' and 'Farmers raise large numbers of sheep and cattle': these being presented as problems, rather than as the basis from which the entire wealth of the country is generated), and is accompanied by a literally Orwellian form for readers to fill in. Pupils are asked to answer leading questions (Would you be prepared to do your bit to reduce greenhouse gases? What does your family think?) and provide personal details that include the names of their teacher, class and school, all to be analysed by government bureaucrats who doubtless maintain a much-consulted database that summarizes the results. The third example, from the UK, uses the time-honoured technique of scare mongering about global warming in order to enlist children to act as 'climate cops' against their parents. The website encourages children to build a 'Climate Crime Case File' for family and friends by watching for crimes which include leaving the tap running, leaving lights on, leaving the TV on standby, failing to use low energy light bulbs, using the clothes dryer on a sunny day and putting hot food in the refrigerator. Though it may not be unreasonable to encourage children to follow these and other similar thrifty habits, it is entirely unacceptable to use the scare of imaginary global warming as a psychological weapon towards that end.

A wide range of other 'educational' websites on climate change (departments of education or environment, councils, science academies, major museums, environmental organizations, and more) provide similar, but often much more subtly concealed, misinformation. The material that these websites present is as skilful and persuasive, as it is offensive to a knowledgeable scientist, and no expense is spared to spread the alarmist message. Add the news media to the mix (discussed next), and it is little wonder that children and well-meaning members of the public – not to mention sophisticated reporters writing for magazines such as *Time*, *National Geographic*, *Scientific American* and *New Scientist* – are duped into believing that a climate apocalypse is at hand.

The message here for all teachers, educators and reporters is that they need to be aware that nearly all climate change information posted on 'official' educational climate websites carries a determined and deliberate propaganda slant.

The role of the media in fostering climate alarmism

The media, both print and electronic, serve to convey to the public the facts and hypotheses of climate change as provided by individual scientists, government and international research agencies and NGO lobby groups. With few exceptions, press reporters commenting on global warming are either uneducated in the science matters involved, or choose to project warming alarm because that fits their personal world view, or are under editorial direction to focus the story around the alarmist headline grab; and often all three. In general, therefore, the media propagate the alarmist cause for global warming, and they have certainly failed to convey to the public both the degree of uncertainty that characterizes climate science and also many of the essential facts that are relevant to human causation of climate change.

Since the turn of the century, it has been a rare day that any metropolitan newspaper failed to carry one or more alarmist stories on climate change, not least because all media proprietors learned long ago that sensational or alarmist news sells best. As one of Australia's most experienced science journalists, Julian Cribb, has remarked[277]:

> The publication of 'bad news' is not a journalistic vice. It's a clear instruction from the market. It's what consumers, on average, demand.... As a newspaper editor I knew, as most editors know, that if you print a lot of good news, people stop buying your paper. Conversely, if you publish the correct mix of doom, gloom and disaster, your circulation swells. I have done the experiment.

Thus climate change hysteria in the media has a life of its own. Ask a web search engine to supply you with references to 'global warming' and it will provide a daily haul of 10-20 alarmist newspaper articles from throughout the world. Many of these stories have as their basis real scientific results from real scientists, but after the results have been processed through public relations staff and compliant media commentators, the result is group-think, political correctness and frisbee science of a high order. A scan through headlines alone, which range from the silly to the ridiculous, will remove any doubt that media treatment of climate change is unbalanced. Reading the articles themselves simply serves to confirm intentional scare mongering and breathtaking scientific ingenuousness.

Alarmist global warming reportage invariably displays one or more of three characteristics. First, it may be concerned with the minutiae of meteorological measurements and trends over the last 150 years in the absence of a proper geological context. Second, it may raise alarm about things that are known to change naturally quite irrespective of human causation, such as ice melting, sea-level change and changes in species' ranges. Third, there is an almost ubiquitous over-reliance on the outputs of unvalidated computer model projections: untestable virtual reality is favoured over actual, real-world data.

On top of such slanted reporting, and in service of the third example just given, the use of careful conditional language has become an invaluable aid for engendering public alarm about global warming. Media writing is infected by a plethora of weasel words such as 'may', 'might', 'perhaps', 'could', 'probably', 'possibly', 'likely', 'expected', 'predicted' and 'modelled'. Journalists and other writers scatter these words through their writings on climate change like confetti. The reason is that – in the absence of hard evidence for damaging human-caused change – assertions about 'possible' change best captures public attention.

Using computer models in support, virtually any type of climatic hazard can be asserted as a possible future change. As one example, a 2005 Queensland State Government report on climate change used these words more than 50 times in 32 pages. That's a rate of almost twice a page. A typical 'could probably' run in this report asserted that Queensland's climate could be more variable and extreme in the future 'with more droughts, heat waves and heavy rainfall' and probably with 'maximum temperatures and heavy downpours ... beyond our current experiences'. Reading further into the report reveals that these statements are all 'climate change projections ... developed from a range of computer-based models of global climate, and scenarios of future global greenhouse gas emissions.' In other words, public policy on this and other science issues is now being set on the outcome of 'Playstation' computer gaming.

Biased coverage by the print media is, however, but part of a much wider problem that involves also radio, television and film coverage of the climate change issue. Chuck Doswell has provided a startling insight into the way that modern 'documentary' films are prepared for cable TV channels – such as the Discovery Science Channel, the History Channel and the National Geographic Channel[278]. As a weather scientist who has participated many times in programmes on severe weather issues, Mr. Doswell comments that the production companies that he has aided invariably:

> ... have the story written before their research even begins. They've decided the 'angle' the story is going to follow, and nothing I say or do seems capable of swaying their determination to produce the story that way. The goal of

the production crew's 'research' ... is to film soundbites ... they can use to back-up the story as it has been written. They are definitely and consistently not seeking to understand the story first on the basis of what they learn by interviewing me. I'm simply there to give credibility to their story.

Mr. Doswell's cynical, but essentially accurate, conclusion is that these types of program – which he terms 'crock-umentaries', or 'disaster porn' – exemplify that:

TV is obviously all about putting eyeballs in front of the advertisements, and has little or nothing to do with public education or offering information to the viewers, whatever pious proclamations they might offer.

This state of affairs is, of course, a reflection of the classic conflict between commercial aims and broadcasting values. The words[279] of British commentator Melanie Phillips regarding the science press in fact apply to nearly all media news and current affairs coverage, which is:

The way global warming is being reported by the science press is a scandal. In selecting only those claims that support a prejudice and disregarding evidence that these claims are false, it is betraying the basic principles of scientific inquiry and has become instead an arm of ideological propaganda.

Finally, however, for all the problems listed above, and much to the outrage of warming alarmists, it should be acknowledged that a handful of quality newspapers have in the past provided a more balanced public discussion of global warming issues. Such papers include the *Wall Street Journal*, the UK's *Daily Telegraph*, the Canadian *National Post*, and *The Australian*. These publications, and a small number of others, have played a vital role in keeping the public informed of both sides of the climate change issue. In the US, *Fox News* television too has recently started to report both sides of the global warming debate. Tellingly, though, the other major US TV channels, and national broadcasters in countries like Australia, New Zealand and Canada, don't even come close to providing equivalently balanced commentary. Meanwhile, that supposed paragon of broadcasting virtue, the British Broadcasting Corporation, employs specialist journalists whose job appears to centre around the deliberate exclusion from news coverage of 'sceptical' views on climate change (see Chapter 12), and the addition of gravitas to the daily dose of shrill global warming alarmism that emanates from environmental and other self-interested groups.

Media science stars: is Hansenism more dangerous than Lysenkoism?

On 23 June 1988, a young and previously unknown NASA computer modeller, James Hansen, appeared before a United States Congressional hearing on climate change. On that occasion, Hansen used a graph (Fig. 33) to convince his listeners that late twentieth century warming was taking place at an accelerated rate, which, it being a scorching summer's day in Washington, a glance out of the window appeared to confirm. He wrote later in justification, in the *Washington Post* (11 February 1989), that 'the evidence for an increasing greenhouse effect is now sufficiently strong that it would have been irresponsible if I had not attempted to alert political leaders'. Hansen's testimony was taken up as a lead news story, and within days the great majority of the American public believed that a climate apocalypse was at hand, and the global warming hare was off and running. Thereby, Hansen became transformed into a media climate star.

Fifteen years later, in 2004, Hansen came to write in the *Scientific American* that 'Emphasis on extreme scenarios may have been appropriate at one time, when the public and decision-makers

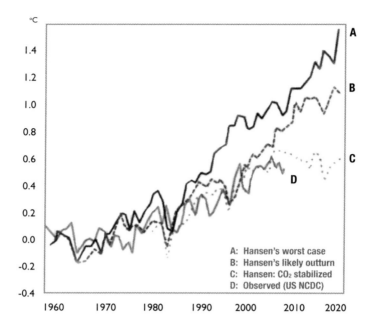

Fig. 33 Jim Hansen's famous computer projections for human greenhouse-driven warming out to 2020, as presented to the US Congress in 1988. Scenarios A and B assume two different levels of continuation of future emissions, Scenario C assumes rapid curtailment of emissions to a stabilized level in 2000. The measured temperature record that has elapsed (D; after the US National Climate Data Center) is plotted for comparison. No cigar for Dr. Hansen.

were relatively unaware of the global warming issue. Now, however, the need is for demonstrably objective climate ... scenarios consistent with what is realistic'[280]. But this conversion to honesty came too late, for in the intervening years thousands of other climate scientists had meanwhile climbed onto the Hansenist funding gravy-train. Currently, global warming alarmism is fuelled by an estimated worldwide expenditure on related research and greenhouse bureaucracy of more than US$10 billion annually. Scientists and bureaucrats being only too human, the power of such sums of money to corrupt not only the politics of greenhouse but even the scientific process itself, should not be underestimated. In recognition of these events, the term Hansenism is now sometimes used to describe the climate hysteria that has gripped Western media sources and political, business and public opinion in a deadly grasp.

Histories of science contain an account of the ideological control of Soviet biology during the mid-twentieth century by plant scientist Trofim Lysenko, who by 1940 had risen to be director of the influential Institute of Genetics of the Russian Academy of Sciences. Lysenko and his supporters rejected the 'dangerous Western concepts' of Mendelian genetics and Darwinian evolution. They preferred the Lamarckian view of the inheritance of acquired characteristics; for instance, that cows could be trained to give more milk and their offspring would then inherit this trait. Whilst this was not an unreasonable hypothesis to erect in the early nineteenth century, by the 1930s the idea had been tested in many ways and was known to be wrong. Requiring its application to agricultural and allied biological research in the USSR was disastrous, not least in the vicious persecution of scientists that took place, and the legacy of this sad episode still disadvantages Soviet biology today. Lysenkoism grew from four main roots[281]:

> a necessity to demonstrate the practical relevance of science to the needs of society;

> the amassing of evidence to show the 'correctness' of the concept as a substitute for causal proof;

> noble cause corruption, whereby data are manipulated to support a cause which is seen as a higher truth;

> ideological zeal, such that dissidents are silenced as 'enemies of the truth'.

Every one of Lysenkoism's four roots is strongly represented in the attitudes of global warming alarmists, and the Hansenist hysteria thereby generated now has a life of its own.

Lysenkoism damaged mainly Soviet science and society, whereas Hansenism has now been exerting a pernicious worldwide influence for more than twenty years. This has already started to undermine the precious public trust from which science traditionally draws its influence and sustenance. It has also damaged the standing of several leading science journals and many science organizations, with the replacement of formerly careful editorial and organizational balance by environmental alarmism and naked global warming advocacy. Future historians of science are therefore likely to judge the 1988-2009 frenzy of climate change alarmism as even more damaging than Lysenkoism, because of the distrust that has now been created about the use of science to inform modern policy-making.

Conclusion

In this chapter we have summarized how the message of global warming alarm is communicated through public and professional meetings, through the education system, and, with reinforcement, through the press. The particular examples given as illustration are but a handful out of a large number of other case histories that could have been used.

To adapt Justice Peter Mahon's memorable phrase during his hearings into the 1987 Mt. Erebus air disaster, towards the end of saving the world from imaginary global warming the public have been subjected to an orchestrated litany of lies about climate change from the very organizations and groups that they trusted and believed in most – including the 'free' press.

9 Science is not about consensus, nor even the IPCC's authority

I regard consensus science as an extremely pernicious development that ought to be stopped cold in its tracks. Historically, the claim of consensus has been the first refuge of scoundrels; it is a way to avoid debate by claiming that the matter is already settled. Whenever you hear the consensus of scientists agrees on something or other, reach for your wallet, because you're being had.

Let's be clear: the work of science has nothing whatever to do with consensus. Consensus is the business of politics. Science, on the contrary, requires only one investigator who happens to be right, which means that he or she has results that are verifiable by reference to the real world. In science consensus is irrelevant. What is relevant is reproducible results. The greatest scientists in history are great precisely because they broke with the consensus.

There is no such thing as consensus science. If it's consensus, it isn't science. If it's science, it isn't consensus. Period.

(Michael Crichton, 2003)[282]

It is a profound truth that science is not concerned with authority, nor with consensus, but with testing hypotheses. Nonetheless, the average citizen, inevitably lacking in advanced understanding of the many fields of modern knowledge, has little choice but to rely upon 'authoritative' sources to form an opinion on most scientific topics. Given, as described in Chapter 7, the corrupt global warming commentary that now pours forth in torrents from scientific societies, research centres and government agencies, invariably with the aid of political spin doctors and a compliantly alarmist press, what are we to do?

The approach taken in the earlier chapters of this book was to explain some of the science issues about global warming in a way that encourages readers to consult relevant sources and make their own assessment. This is, of course, the traditional scientific method, which relies on observations, established laws and simple logic, and pays no attention to consensus or authority. Regrettably, though, the global warming discussion in the public domain ceased to be about science many years ago, ever since the coining and relentless reiteration of the silly mantras that 'the science is settled' and 'there is a consensus'. As the December 2009, Copenhagen conference confirmed (Chapter 12), the climate change issue is now primarily about very big politics and very big business.

Given their political aims, therefore, it is no surprise that most of today's public commentators on global warming stress the *authority* of the climate pronouncements made by the Intergovernmental Panel On Climate Change (IPCC) and its supporting organizations. Often added too is the vacuous claim that a *consensus* of scientists agrees with the IPCC views, as if that were somehow scientifically relevant even were it to be true. At the same time, continual unsolicited *ad hominem* attacks are made on qualified persons who espouse different views, and who are often disparaged as 'sceptics', 'deniers', or worse.

Having dealt with most of the other authority figures in Chapters 7 and 8, we will now turn to an examination of the credentials of the IPCC, followed by a summary of credible sources of alternative advice, and a quick lesson in the techniques that are used to try to discredit those who provide such alternative viewpoints.

The Intergovernmental Panel on Climate Change (IPCC)

Two things need to be understood at the outset, in the face of dominant counterveiling propaganda. The first is that the IPCC does not represent the views of thousands of scientists, but of a narrowly selected few hundred at most, most of the other participants being social scientists or government bureaucrats; indeed, only 51 persons participated as authors in the preparation of the Summary for Policymakers for the science volume of the Fourth Assessment Report, and not even all of those are practising research scientists. The second is that despite being advised by scientists, among other persons, the IPCC overall is a profoundly political and not a scientific body. These matters, and the resulting scientific inadequacies of the way that IPCC handles the scientific issues relating to climate change, especially in its assessment reports, are explained in the discussion below.

Background

The IPCC was established in 1988 by the United Nations under the co-sponsorship of the World Meteorological Organization (WMO) and the United Nations Environmental Program (UNEP), with a Charter that directs it to assess peer-reviewed research 'relevant to the understanding of the risk of human-induced climate change'. The terms of reference of the IPCC include:

> assessing the evidence on climate change and its impacts

> assessing the options for adapting to or mitigating climate change

> providing advice, both scientific and socio-economic, to the United Nations Framework Convention on Climate Change (UNFCCC)

Based on the participation of professional advisers, many of whom are government-appointed scientists and bureaucrats, the IPCC produces comprehensive assessment reports of the presumed human greenhouse influence on today's climate. To date, these reports are the First (1990), Second (1996), Third (2001) and Fourth (2007) Assessment reports[5], with a Fifth due in 2013-2015. These reports have become increasingly complex and difficult to read. For example, the Third Assessment Report (3AR) involved 122 lead authors, 515 other contributors, 21 review editors and about 700 reviewers, and the resulting 881-page document took three years to produce.

The IPCC operates in close relationship with the United Nation's Framework Convention on Climate Change (FCCC), which, you may remember from Chapter 1, defines climate change as 'a change of climate which is attributed directly or indirectly to human activity that alters the composition of the global atmosphere and which is in addition to natural climate variability observed over comparable time periods.' Thus at its point of origin and reporting, the IPCC is set up to consider not climate change in general, but only change caused by human perturbation of the atmosphere.

The First Assessment Report of the IPCC was published in 1990, and formed the science basis for the 1992 Rio Climate Summit that in turn led to the Framework Convention on Climate Change and the 1997 Kyoto Protocol. Thus, as intended by its founders, advice from the IPCC is the linear thread that underlies all national and international efforts to control the emission of greenhouse gases. This unbalanced brief inevitably leads to advocacy research[283], and the provision of dangerously inadequate advice.

A new assessment report is produced by each of IPCC's three major working groups about every five years, the groups being: Working Group 1 – Physical Science Basis; Working Group 2 – Impacts, Adaptation and Vulnerability; Working Group 3 – Mitigation of Climate Change ('mitigation' being IPCC-speak for 'prevention', for you can't mitigate something which hasn't started yet). Overall, most IPCC authors are therefore not scientists, but rather geographers, sociologists and economists, as a consequence of which the IPCC is mostly about advocating social policy rather than conducting dispassionate science.

The Summary for Policymakers (SPM)

The Working Group 1 (WG1: Science) reports of the IPCC summarize a large amount of good climate science, to which many scientists have contributed in good faith. The problem, in general, does not lie with these lengthy assessment reports, which are anyway not read by politicians, policy-makers or the general public. Rather, non-experts rely for their information on the brief, pre-digested Summary for Policymakers (SPM) which accompanies the WG1 report.

The SPM is a considered statement deliberately shaped to be suitable for use by government policymakers. Each SPM is compiled by a panel of senior scientists, after which it is then edited and approved line by line by a 400-strong committee of Government-appointed delegates from participating countries, many of whom are not necessarily working scientists. By this process, the uncertainties in our understanding of climate change become played down, and the likelihood of overemphasizing an anthropogenic cause for climate change is increased.

Though based upon recommendations from qualified scientists, and resting on its parent assessment report, the science SPM in fact carries a heavy political overlay. Indeed, in 2007 the main Fourth Assessment Report (4AR) science report had to be edited after the event to bring it into harmony with what had been agreed among governmental representatives for the SPM. Such a reversal of normal science procedure does not inspire confidence in the accuracy of either document.

Because governments that have signed the Framework convention accept the SPM as the basis for setting their climate policy, the IPCC, through this document, in effect acts as a monopoly provider of advice on climate change which, in general, is not subjected to further due diligence auditing[284].

What is the basis for alarm?

The four successive IPCC science assessment reports provide a detailed treatment of many aspects of climate science as was reflected in the refereed scientific literature at the time they were written, and they serve as valuable source compilations of published research. At the same time, the heavy overemphasis that is placed on deterministic computer modelling, and the persistent downplaying of natural climate variation, means that the IPCC reports fail to provide a balanced summary of climate change 'in the round'.

Each successive IPCC science report has incrementally talked up the threat of dangerous human-caused change, especially in the influential SPMs, as follows:

- The observed [twentieth century temperature] increase could be largely due to ... natural variability (1AR, 1990);
- The balance of the evidence suggests a discernible human influence on climate (2AR, 1996);
- There is new and stronger evidence that most of the warming observed over the last 50 years is attributable to human activities (3AR, 2001);
- Most of the observed increase in globally averaged temperature since the mid-twentieth century is very likely [= 90 per cent probable] due to the observed increase in anthropogenic greenhouse gas concentrations (4AR, 2007).

What can the evidence have been for these increasingly dramatic warnings, for it is certainly hard to find convincing arguments in the assessment reports. Ironically, the reports were issued over the same time period that evidence published by independent scientists has incrementally undercut every one of the main IPCC arguments for asserting a dangerous human influence on climate (Chapter 1, the quiet transition between solar cycles 23 and 24; Chapter 2, the post-1998 temperature cooling, despite continuing carbon dioxide increase; Chapter 3, the weak temperature effect of increasing carbon dioxide; Chapter 4, the lack of acceleration of sea-level rise; Chapter 5, statistical climate models project twenty-first century cooling; Chapter 6, circumstantial evidence of warming is not evidence of its human cause; and Chapter 7, the malfeasant hockey-stick). Note, in particular, that none of the four celebrated IPCC statements above say directly that a human global warming influence has been identified and measured, for the very good reason that it hasn't.

A balanced summary statement that might have been made in 1996, and that remains true today, could have read: 'There is unanimous agreement that human activities can affect climate at local scale; summed across the globe, these local affects may have a measurable effect on global climate; for the period of the instrumental record (say the last 100 years), however, climate change has proceeded at rates that lie within previous natural rates and magnitudes, and any anthropogenic effect cannot be distinguished from the noise and natural variation in the system.' One can only wonder why the IPCC did not make such a statement in its most recent report.

IPCC's claimed 'gold standard' of peer review
It is often claimed that the authority of IPCC's policy advice rests upon the fact that its scientists use only information which has been published in reputable, refereed journals, and that their science summaries are also subjected to high standards of peer review. Indeed, with astonishing chutzpah given the torrent of criticism that they are currently enduring for including in the 4AR an unchecked, anecdotal statement that the Himalayan icecap is melting at dangerous rates, senior IPCC figures have recently been asserting that their reports set the gold standard of reviewing procedure. Alas, their review procedures in fact fall far short of acceptable scientific practice[8], as exemplified yet again recently by the disclosure that the 2007 4AR contains at least 20 references

(and counting) to reports and papers commissioned by environmental advocacy groups, including Greenpeace and the World Wildlife Fund[285]. This reduces the IPCC reports to the status of the 'grey' promotional literature that environmental NGOs commission and deploy so effectively. Not that this is a particularly new thought, for the IPCC has much historical form for indulging in questionable scientific practice[286].

Though the IPCC does seek outside reviewers, the editing of IPCC technical volumes is performed under the control of trusted lead authors who dismiss criticism of the presumption of a dangerous human influence on global climate. The tight IPCC editing group selectively favours the citation of papers that claim man-made warming, and the lead authors also have a record of seeking advice from within a narrow coterie of like-minded scientists, and of ignoring criticism from independent reviewers. A stunning example is provided by the editorial rejection of the following criticism of the pivotal Chapter 9 (*Understanding and Attributing Climate Change*) in the 4AR. The comments[287] below were made by a senior scientist, who works for the NASA group that is responsible for the Goddard Institute of Space Studies (GISS) temperature curve:

> There is no scientific merit to be found in the Executive Summary. The presentation sounds like something put together by Greenpeace activists and their legal department. The points being made are made arbitrarily with legal sounding caveats without having established any foundation or basis in fact. The Executive Summary seems to be a political statement that is only designed to annoy greenhouse sceptics. Wasn't the IPCC Assessment Report intended to be a scientific document that would merit solid backing from the climate science community – instead of forcing many climate scientists into having to agree with greenhouse sceptic criticisms that this is indeed a report with a clear and obvious political agenda. Attribution

cannot happen until understanding has been clearly demonstrated. Once the facts of climate change have been established and understood, attribution will become selfevident to all. The Executive Summary as it stands is beyond redemption and should simply be deleted.

Following on from that, the overall inadequacy of IPCC reviewing procedures has been well summarized by Professor von Storch, who wrote in 2005[288]:

> The IPCC has failed to ensure that the assessment reports, which shall review the existing published knowledge and knowledge claims, should have been prepared by scientists not significantly involved in the research themselves. Instead, the IPCC has chosen to invite scientists who dominate the debate about the considered issues to participate in the assessment. This was already in the Second Assessment Report a contested problem, and the IPCC would have done better in inviting other, considerably more independent scientists for this task. Instead, the IPCC has asked scientists like Professor Mann [of hockey-stick fame] to review his own work. This does not represent an 'independent' review.

This discussion notwithstanding, even best-practice peer review plays a mostly editorial rather than a scientific quality assurance role, and many persons, including senior scientists, overrate its importance. For example, in January 2007, the CEO of the UK's National Environmental Research Council (NERC) engaged in a public discussion of the global warming issue. In defending the IPCC's reliance on 'gold standard' (his term) peer-reviewed literature, Professor Thorpe had this to say: 'How can one judge the scientific credibility of a statement from a document on a given website unless it can be shown to have at least been through the peer review process?'

The answer to this question, of course, depends entirely upon who 'one' is. If one is a member of the general public, or a politician, then Thorpe's advice that you should check to see if the work has been peer reviewed is not unreasonable. But if one is instead a trained scientist, then the answer is different: it is that you should assess the validity of the matter by processing the scientific arguments through your own brain, or, failing a capacity to do that, simply make no comment – for that a piece of science has been peer reviewed, or is recommended by an authority, has never been, and will never be, an acceptable means of adjudicating a scientific issue.

Other inadequate procedures

Many other severe and unrebutted criticisms have been made of the procedures used by the IPCC in the preparation of its assessment reports. However, until very recently[289] the organization has just ignored them and cruised along on its majestic path of saving the planet, and of course our grandchildren, from imaginary climate disaster.

Though many scientists have expressed their discontent, including some who are or were participants in IPCC procedures, the most detailed and accurate analysis of the IPCC's shortcomings has been undertaken by John McLean, a climate data analyst from Melbourne. McLean's auditing reports, referred to earlier[8], were made using data released by the IPCC, and each report explains fully the way in which the data has been analysed in order to reach the conclusions that are drawn. Deficiences of IPCC procedure are known to be:

- The IPCC promulgates misleading statistics about those who participate in its deliberations. For example, it is claimed that 'more than 2500 scientists' have participated in or approved the recommendations in the IPCC's 4AR. In actuality, just 51 persons participated in the final approval of the policymakers' summary for the Working Group 1 science report.

In an even more astonishing example, McLean (2008) reports

that out of the 62 expert reviewers of the critical 4AR Chapter 9, *Understanding and Attributing Climate Change*, 55 had a conflicting or vested interest, leaving only seven reviewers who can be viewed, *prima facie*, as impartial[8]. Seven, of course, is a very different number from 'more than 2500'. And the chapter itself was written in the first place by a small, interlinked group of authors, many of whom are employed at only three climate change institutions from among hundreds of qualified persons and organizations worldwide. This is the very same chapter for which the executive summary was excoriated (quotation, p. 198) as having 'no scientific merit' and being a 'political statement'.

• Meritorious scientists participating in IPCC activities have expressed their dissatisfaction at what they perceive as political interference in the preparation of IPCC reports. For example, Dr Chris Landsea, an acknowledged expert on hurricanes, withdrew his participation in the IPCC in 2005 with the comment that: 'I personally cannot in good faith continue to contribute to a process that I view as both being motivated by pre-conceived agendas and being scientifically unsound.'[290]

• In order to make projections of future climate change, the IPCC developed in 3AR a number of alternative socio-economic scenarios for future world energy use. These scenarios, which feed into all climate projections made by the IPCC, have been shown to be both unrealistic and flawed[111]. Calculations concerning per capita income, economic growth and greenhouse gas emissions in different regions therefore substantially overstate the likely growth in output of developing countries during the twenty-first century. In particular, the IPCC converts national Gross Domestic Profit (GDP) data to a common measure using market exchange rates rather than the now more-widely used purchasing power parity rates. Despite their known inaccuracy, these calculations were adopted unchanged for the climate projections made in the 4AR, and appear to be on track for use also in the Fifth report, due in 2013.

- In 3AR and 4AR, the IPCC adopted a qualitative scale of probability terms that has no rigorous basis. Terms such as 'likely' (>66 per cent probable) and 'very likely' (>90 per cent probable) have no actual statistical meaning but instead represent only considered opinions. This terminology is, at best, highly misleading and represents sociology, not science. The regrettable practice of allocating scientifically meaningless numerical probabilities is also practiced in the Summary for Policymakers. Such chicanery has no place in what is supposed to be a science-based public policy document.

Having learned from witnesses about these and other deficiencies in IPCC procedures, in 2005 the authoritative UK House of Lords Select Committee on Economic Affairs concluded that[291]:

> 'We can see no justification for an IPCC procedure which strikes us as opening the way for climate science and economics to be determined, at least in part, by political requirements rather than by evidence. Sound science cannot emerge from an unsound process.'

Since the IPCC's 3AR in 2001, governments, business, the media and environmental activists have all assumed that conclusive evidence has been provided that human greenhouse gas emissions are causing dangerous global warming. This public view is a result of IPCC's superb and no-costs-spared marketing. In fact, the IPCC has provided nothing of the sort, but rather has a predisposition towards that conclusion built into its terms of reference. Most citizens remain blissfully unaware of the flawed nature of IPCC reports and advice, or, if the matter comes to their attention, they simply shrug and repeat one or another of the meaningless mantras of the Green movement, such as 'well, I want to do the right thing by the planet.'

However, and despite the overwhelming public assumption of human causation, the 3AR of the IPCC actually said (2001, p. 97) that:

> The fact that the global mean temperature has increased since the late nineteenth century and that other trends have been observed does not necessarily mean that an anthropogenic effect on the climate has been identified. Climate has always varied on all time-scales, so the observed change may be natural.

This statement remains true today; indeed, it is the very kernel of the null hypothesis (Chapter 6).

Alternative sources of advice

Thanks not a little to the remorseless barrage of *ad hominem* abuse directed at independent scientists (see next section), until recently[292] the public has also remained mostly unaware that very large numbers of highly qualified, independent scientists have expressed non-alarmist views on global warming that run completely counter to the IPCC's opinions. Several websites contain useful lists of documents, declarations and letters signed by professional persons of standing, many of whom are also qualified in one or another field of science or climate science[293].

Readers are encouraged to consult these documents directly themselves, but I will give a brief summary of some of the most important here.

First is an important single-source scientific compendium that carries the same weight with independent scientists as the IPCC assessment reports do with IPCC supporters. Edited by Craig Idso and Fred Singer, who were assisted by a panel of 22 other climate scientists, the report of the Non-Intergovernmental Panel on Climate Change (NIPCC)[112] is a *tour de force*. It summarizes many important papers that are missing from the IPCC reports, and contains a measured appraisal of most of the major, controversial topics of climate science. Though a total of only 24 scientists participated, science progresses by testing hypotheses, not by counting heads. Throughout the NIPCC report are scattered many independent tests of the hypothesis of the day – which is that human-sourced carbon dioxide emissions are causing dangerous warming – and the hypothesis fails nearly all of them. Another similarly balanced though less detailed alternative to the IPCC's SPM is an excellent summary document produced by the Canadian Fraser Institute[294].

Moving on to shorter statements that carry a larger number of signatories, two of the most important were letters written to the secretary general of the United Nations at the time of the Bali (December 2007) and Copenhagen (December 2009) climate conferences.

The Bali letter[295] contained the following statements:

> Recent observations of phenomena such as glacial retreats, sea-level rise and the migration of temperature-sensitive species are not evidence for abnormal climate change, for none of these changes has been shown to lie outside the bounds of known natural variability.

> The average rate of warming of 0.1-0.2°C/decade recorded by satellites during the late twentieth century falls within known natural rates of warming and cooling over the last 10,000 years.

> Leading scientists, including some senior IPCC representatives, acknowledge that today's computer models cannot predict climate. Consistent with this, and despite computer projections of temperature rises, there has been no net global warming since 1998. That the current temperature plateau follows a late twentieth century period of warming is consistent with the continuation today of natural multi-decadal or millennial climate cycling.

The distinguished list of 103 signatories to this letter included many winners of awards, medals and prizes in meteorology, climatology or cognate subdisciplines, and 24 are Emeritus Professors.

The 2009 Copenhagen letter[296] was accompanied by a similar but longer (143) list of signatories, and included the comment:

> Climate change science is in a period of 'negative discovery' – the more we learn about this exceptionally complex and rapidly evolving field the more we realize how little we know. Truly, the science is NOT settled.
>
> Therefore, there is no sound reason to impose expensive and restrictive public policy decisions on the peoples of the world without first providing convincing evidence that human activities are causing dangerous climate change beyond that resulting from natural causes. Before any precipitate action is taken, we must have solid observational data demonstrating that recent changes in climate differ substantially from changes observed in the past and are well in excess of normal variations caused by solar cycles, ocean currents, weather cycles (El Nino, etc.), changes in the Earth's orbital parameters and other natural phenomena.

A third recent public statement on climate change, the Manhattan Declaration, was first declared at a Climate Change meeting in New York in March 2008[297], and concluded:

> That current plans to restrict anthropogenic CO_2 emissions are a dangerous misallocation of intellectual capital and resources that should be dedicated to solving humanity's real and serious problems.
>
> That there is no convincing evidence that CO_2 emissions from modern industrial activity has in the past, is now, or will in the future cause catastrophic climate change.

That attempts by governments to inflict taxes and costly regulations on industry and individual citizens with the aim of reducing emissions of CO_2 will pointlessly curtail the prosperity of the West and progress of developing nations without affecting climate.

At the end of 2009, the Manhattan Declaration had attracted 1,307 signatories, of whom 707 are persons with professional qualifications in a science or policy-related area.

Finally, among several other, similar statements or declarations, perhaps the two most important are by the 650 scientists listed in a 2009 US Senate report (Morano, 2009)[298], and the Oregon Petition[299], in which more than 30,000 persons, 9,029 with PhD degrees, put their names to the following statement:

There is no convincing scientific evidence that human release of carbon dioxide, methane, or other greenhouse gasses is causing or will, in the foreseeable future, cause catastrophic heating of the Earth's atmosphere and disruption of the Earth's climate. Moreover, there is substantial scientific evidence that increases in atmospheric carbon dioxide produce many beneficial effects upon the natural plant and animal environments of the Earth.

Given that the Oregon Petition was first promulgated in 1998, perhaps the most surprising thing about it, and the other later statements referred to above, is how little the basic message has changed over recent years. It has in fact been crystal clear to most independent scientists since before the turn of the century that no observational basis exists for global warming alarmism.

Scientists associated with the IPCC are bound by a 'cabinet solidarity' principle to the politically nuanced advice that is contained in the IPCC's Summary for Policymakers. In contrast, the signatories of the letters and submissions referred to above, like the authors of the research papers referred to throughout this book, provide their judgements independent of anything other than scientific consideration. Their prosaic, non-alarmist conclusions about climate change are, or course, of little value in helping politicians to build careers or media organizations sell advertising, which is presumably why they are largely ignored by the press.

Science truth is not determined by head counts. Nonetheless, there is now overwhelming, documented evidence that a large number of responsible, highly qualified professional scientists, engineers and economists do not accept that the advice given by the IPCC is accurate or wise enough to be relied upon for setting climate policies. The considered views of such a large body of expert people cannot simply be wished away.

Confusing the debate with rhetoric

Independent scientists are challenging the advice of the IPCC when they raise doubt about the legitimacy of a particular piece of climate alarmism – say that Tuvalu is being swamped by a rising sea-level. Any response to such criticism that reaches the press almost never deals with the science issue in question. Instead, simple but highly effective rhetorical devices are used to counter the doubts or to challenge the integrity of those who raise them.

The following irrelevant assertions are very commonly used by climate alarmists, through the media, to negate sensible public discussion of global warming, and especially to counter the views of independent scientists.

- 'The science is settled' or, 'there is a consensus on the issue'. In reality, science is about facts, experiments and testing hypotheses, not consensus; therefore, science is never 'settled'.

- 'He is paid by the fossil fuel industry, and is merely repeating their spin.' An idea is not responsible for those who believe in it, and neither is the validity of a scientific hypothesis determined by the character or beliefs of the persons who fund or undertake research to test it. Science discussions are determined on their merits, by using tests against empirical or experimental data. It may be hard to believe in a post-modern world, but who paid for the data to be gathered and assessed is simply irrelevant.

- *'She works for a left wing/right wing think tank, so her work is tainted.'* Think tanks serve an invaluable function in our society. On all sides of politics they are the source of much excellent policy analysis. They provide extended discussion and commentary on matters of public interest, and have made many fine contributions towards balancing the public debate on climate change. That all think tanks receive funding from industry sources is an indication that those that survive are delivering value for money, and does not impugn their integrity.

- *'He is just a climate sceptic', a 'contrarian', a 'denialist'.* These terms are used routinely to denigrate independent scientists. The first two are amusingly silly. First, because most people termed climate 'sceptics' are in fact climate 'agnostics', and have no particular axe to grind regarding human influence on climate; rather, they just want the facts to be established, and to fall where they may. Second, because all good scientists are sceptics: that is their professional job. To not be a sceptic of the hypothesis that you are testing is the rudest of scientific errors, for it means that you are committed to a particular outcome: that's faith, not science. Lastly, introduction of the term 'denialist' into the public climate debate, with its deliberate connotations with holocaust denial, serves only to cheapen those who practice the custom.

- *'Six Nobel Prize winners, and seven members of the National Academy say ...'* Argument from authority is the very antithesis of the scientific method. That the Royal Society of London tried to restrict the public debate on climate change through intimidation of Esso UK (Chapter 7), for example, was a complete betrayal of all that the Society, and the scientific method, stands for.

- *'The "precautionary principle" says that we should limit human carbon dioxide emissions because of the risk that the emissions will cause dangerous warming'.* As explained next in Chapter 10, the precautionary principle is valueless for setting public policy. This is because the 'principle' says nothing about the relative likelihood of particular risks, including the risks of taking a precautionary action as well as the risks of not taking it. As we all know, life is full of risks. We only have the time and money to take precautions against those that have a significant degree of likelihood, and for which the risks of precautionary action are demonstrably less than the risks of taking no action.

The rhetorical devices listed above are used to combat the views of independent, non-alarmist scientists, and are often supplemented by two other techniques. The first is the maintenance of trashy, though impressive-looking, websites that specialize in providing prejudicial and often inaccurate information about well-known independent scientists; typical examples of the genre include SourceWatch, ExxonSecrets and DeSmogBlog. The second is the widespread activity of blog 'trolls', who monitor a huge number of websites while ever alert to favourable mentions of non-alarmist scientists. Such mentions, when found, are counteracted by various types of *ad hominem* criticism, often drawn from the three websites named above, accompanied also by one or more of the six listed rhetorical devices.

Conclusion

An inspection of daily media writings about global warming quickly reveals that whenever the views of a 'climate sceptic' are discussed, one or more of six *ad hominem* arguments are used in rebuttal. Almost by definition, one can say that the persons who use these and other authority-style arguments are either ill educated, or alternatively so well-educated that they know that they will lose any discussion that is allowed to take place on science grounds.

It is for this same reason that IPCC scientists are so reluctant to appear in formal debate against other well qualified, independent scientists. As many who have tried have found out, a persistent and intransigent refusal by IPCC scientists to debate their critics in public is the main reason why conferences do not occur at which equal numbers of scientists, and equal time, are allocated to both main sides of the debate. After all, when you hold the high ground of having institutional science, national government and United Nations support, as do the climate alarmists, why would you demean yourself by debating your opponents in public; for you have everything to lose and nothing to gain.

10 The cost of precaution

> We can confirm our initial view that the term
> 'precautionary principle' should not be used, and
> recommend that it cease to be included in policy guidance
> … In our view, the terms 'precautionary principle' and
> 'precautionary approach' in isolation from any such
> clarification have been the subject of such confusion and
> different interpretations as to be devalued and of little
> practical help, particularly in public debate.
>
> (UK House of Commons Select Committee on Science and
> Technology, 2006)[300]

Having read this far, you might conclude that the message of this book is: 'human-caused global climate change is so small that it cannot be measured, so just relax and forget about climate change altogether.' Not so fast; for though the first part of that statement is true, the second is far from sensible. The reason is that the greatest damage that has been inflicted by those whipping up the hypothetical threat of human-caused global warming is that the subsequent hysteria has overwhelmed mature consideration of the much greater and proven threat of *natural* climate change.

There is irony as well as irresponsibility here. It is that the type of cooling (not warming, mark) that the globe has experienced since around the turn of the twenty-first century, and which some leading scientists predict will become significantly more intense because of our currently quiet sun, is by its very nature more threatening to human activity and health than an equivalent amount of benign warming would be. The prescribed remedy for those who might be inclined to doubt this is a study course in the history of the Little Ice Age, the various phases of which extended from the late thirteenth to the mid-nineteenth century[301].

Giving Earth the benefit of the doubt

A common expression of human caution, espoused by influential media proprietor Rupert Murdoch[302], is that in matters of potentially dangerous human-caused global warming we should 'give Earth the benefit of the doubt'. Such a slogan bears all the hallmarks of having been produced by a green advertising agency. The catchy phrase reveals a profound misunderstanding of the real climatic risks faced by our societies, because it assumes that global warming is more dangerous, or more to be feared, than is global cooling. In reality, precisely the opposite is true.

The precautionary principle

This 'principle', which is sociological and not scientific, was introduced in order to assist governments and peoples with risk analysis of environmental issues. First formulated at a United Nations environment conference at Rio de Janiero in 1992, it stated that: 'Where there are threats of serious or irreversible damage, lack of full scientific certainty shall not be used as a reason for postponing cost-effective measures to prevent environmental degradation.'

Faced as they are with a lack of compelling science on their side, many global warming devotees invoke the precautionary principle as a means of forcing action against what they feel, but cannot show, is a risk of dangerous human-caused warming. Indeed, the very introduction of the precautionary principle into the argument in the first place is an acknowledgement that no compelling scientific evidence for alarm exists.

Often, too, the precautionary principle represents a moral precept masquerading under a scientific cloak. Adhering to moral principle through thick and thin is certainly a part of the precautionary principle as practiced by many environmentalists; it is a principle of the wrong type to be used for the formulation of effective public environmental policy, which instead needs to be rooted in evidence-based science. Scientific principles acknowledge the supremacy of experiment and observation, and do not bow to untestable moral propositions.

Despite a disturbing lack of intellectual rigor, not to mention the presence of ambiguity in the original and other definitions, the precautionary principle has been incorporated into law in several countries. For instance, in Australia, the Commonwealth Fisheries Management Act 1991 (Section 516A) requires the regulatory authority 'to pursue the objective of ensuring that the exploitation of fisheries resources and the carrying on of any related activities are conducted in a manner consistent with the principles of ecologically sustainable development and the exercise of the precautionary principle'. Experience shows, too, that the adoption of the precautionary principle as even a policy guideline is inevitably followed later by the development of legally binding precautionary rules[303].

What is precaution to be taken against?

In any case, even should we wish to take precautions, we have to know what to take them against. As we have seen in Chapter 5, the IPCC's deterministic computer models project that the global temperature in ten years time will be warmer than today's. Other, statistical, computer models, based upon the projection of past climate patterns, indicate that global temperature will be cooler ten years hence. The reality is, therefore, that no scientist can tell you with confidence whether the temperature in 2020, let alone 2100, will be warmer or cooler than today's.

In the face of this situation, those who nonetheless wish to apply the precautionary principle to near-future climate change need to reflect on the strong likelihood that significant, and perhaps damaging, global cooling will eventuate over the next few decades. Any such cooling will have a strong negative impact on the major mid-latitude grain-producing areas of the northern hemisphere. In such circumstances, the precautionary thing to do would be to increase the amount of carbon dioxide in the atmosphere because of its mild warming and plant fertilization effects. At the same time, given its value as a non-renewable energy source, it would also be a sensible precaution not to irresponsibly squander the extra 30-40 per cent of coal that is required to sequester the carbon dioxide emitted by coal-fired power plants, at the same time thereby happily avoiding the great costs that are engendered for no measurable environmental benefit should a sequestration policy be implemented.

The only sensible precaution that can be taken in our present circumstances is to plan for a continuation of the present climate trend, and to recognize and plan also for reasonable bounds of future climate variability. As the temperature trend for ten years now has been one of cooling, since the unusually warm El Nino year of 1998, this requires an initial precautionary response to cooling rather than warming.

In reality, though, it is not soppy, feel-good precaution that is required to protect our citizens and environment, but hard-nosed and effective prudence. It is absolutely clear from, for example, the 2005 Hurricane Katrina disaster in the US, the 2007 floods in the United Kingdom and the tragic bushfires in Australia in 2009, that the governments of even advanced, wealthy countries are inadequately prepared for climate-related disasters of natural origin. Undoubtedly, one of the reasons for this is that the self-same governments have been distracted by the hysterical fuss created by the Greens and other self-interested groups about entirely hypothetical and yet-to-be-measured human-caused global warming.

The cost of precaution

Those who propose the curtailment of human carbon dioxide emissions as a precaution against future warming invariably fail to address two key cost:benefit issues.

First, it is an iron fact that observably different climates exist around the planet at any one time, and it is equally undeniable that climate everywhere varies through time. Against this background, there is simply no credible published research that shows that the net human costs and environmental risks of a given amount of future global warming exceed the costs and risks of an equivalent global cooling. Indeed, the complexity of Earth's regionally different climates, overlain with the further complexity of nation state politics, precludes any sensible global approach to this issue, and such cost:benefit analyses are probably only possible on a limited scale for small, individual nations.

The second cost:benefit issue that is rarely confronted is the question of what actual temperature benefit (reflexively presumed to have the character of 'warming prevented') will result from proposed emissions reductions schemes, whose costs are certain to be in the swingeing trillions-of-dollars range.

Presuming that all signatory nations meet their obligations under the Kyoto Protocol (which isn't going to happen, though that's another story), modelling suggests that $0.2°C$ of warming might be prevented by 2100[304]. Beyond that, in discussions about the mooted carbon dioxide cap-and-trade bill in the US, and other international carbon dioxide trading plans for the Copenhagen conference, climatologist John Christy gave the following evidence to the US Congress Ways and Means Committee[305]:

> I calculated using IPCC climate models that even if the entire country [USA] adopts this rule, the net global impact would be at most one hundredth of a degree by 2100, and even if the entire world did the same, the effect would be less than four hundredths of a degree by 2100, an amount so tiny we cannot measure it with instruments or notice it in any way.

That such unrealistic, indeed entirely stupid, 'precautionary' policies are being seriously contemplated by world political leaders in 2009 is a devastating indictment of the modern political process. The situation represents the triumph of politically cynical green opportunism and allied commercial and other self-interest[306], and results from a costly, professional and all-encompassing propaganda and advertising campaign that is without parallel in history. The campaign has been spearheaded by large environmental NGOs in concert with governments, and Vaclav Klaus (president of the Czech Republic) stands almost alone among modern heads of state in discerning the threat that it poses, viz[307]:

> As someone who lived under communism for most of my life I feel obliged to say that the biggest threat to freedom, democracy, the market economy and prosperity at the beginning of the 21st century is not communism or its various softer variants. Communism (has been) replaced by the threat of ambitious environmentalism ....

> The environmentalists consider their ideas and arguments to be an undisputable truth and use sophisticated methods of media manipulation and PR campaigns to exert pressure on policymakers to achieve their goals. Their argumentation is based on the spreading of fear and panic by declaring the future of the world to be under serious threat. In such an atmosphere they continue pushing policymakers to adopt illiberal measures, impose arbitrary limits, regulations, prohibitions, and restrictions on everyday human activities and make people subject to omnipotent bureaucratic decision-making ....

> Manmade climate change has become one of the most dangerous arguments aimed at distorting human efforts and public policies in the whole world.

Prudence is better than precaution

Independent scientists who have considered the matter carefully do not deny that human activities can have an effect on local climate, nor that the sum of such local effects represents a hypothetical global signal. The key questions to be answered, however, are, first, can any human global signal be measured, and, second, if so does it represent, or is it likely to become, dangerous change outside of the range of natural variability?

The answers to these questions were developed earlier in this book. My conclusions are that no human global climate signal has yet been measured, and it is therefore probable that any such signal lies embedded within the variability of the natural climate system. Meanwhile, global temperature change is occurring, as it always naturally does, and a phase of cooling has succeeded the mild late twentieth century warming.

It is certain that natural climate change will continue in the future as it has in the past – including warmings, coolings and step events. In the face of this, it is clearly most prudent to adopt a climate policy of preparation for, and adaptation to, climate change as and when it occurs. Adaptive planning for future climate events and change, then, should be tailored to provide responses to the known rates, magnitudes and risks of natural change. Once in place, these same plans will provide an adequate response to any human-caused change should it emerge in measurable quantity at some future date.

Conclusion

Despite its adoption during the 1990s as a basis for law-making in several Western countries, the Science and Technology Committee of the UK House of Commons has come to the firm conclusion quoted at the head of this chapter; namely that the principle has no value for use in policy guidance. Regarding climate change in general, rather than human-caused global warming in particular, what is needed instead is a prudent policy of preparation for, and response to, all climatic events and hazards as and when they develop. This theme of adaptation to climate change will be explored further in the next, penultimate chapter.

11 Plan B: a fresh approach

> I know that most men, including those at ease with
> problems of the greatest complexity, can seldom accept
> even the simplest and most obvious truth if it be such as
> would oblige them to admit the falsity of conclusions
> which they have delighted in explaining to colleagues,
> which they have proudly taught to others, and which they
> have woven, thread by thread, into the fabric of their lives.
>
> (Leo Tolstoy)

To say that human-caused global warming is proven to be a
dangerous problem is untrue, and to introduce futile policies
aimed at 'stopping climate change' is both vainglorious and hugely
expensive. Nonetheless, and despite the failure of the hypothesis of
dangerous human-caused global warming, all studies of ancient
climate indicate that a very real climate problem does exist. It is the
risk associated with *natural climatic phenomena*, including short-term
events such as floods and cyclones, intermediate scale events such
as drought, and longer term warming and cooling trends.

In dealing with the certainties and uncertainties of these
demonstrable natural events, and at the same time allowing for
possible human-caused climate change, the key issue is prudent risk
assessment. The main certainty is that natural climate change is
going to continue, and that some manifestations – droughts, storms
and sea-level change, for example – pose dangers that will be
expensive to adapt to. In contrast, the great danger posed by
current global warming hysteria is that it distracts attention and
resources away from the development of sound policies of

adaptation to the natural climate vicissitudes that are a certain part of our future.

The real risk is natural climate change

Study of the geological record (Chapter 1) reveals many instances of natural climate change of a speed and magnitude that would be hazardous to human life and economic wellbeing were they to be revisited upon today's planet[308].

One famous example of this is the Younger Dryas climatic event, which marked a dramatic cooling to near-glacial conditions during the general post-glacial warming. The event is well recorded in the Greenland ice core (Fig. 4), and is characterized by several rapid climatic changes. For example, the warmings that preceded and ended the Younger Dryas occurred in three years and 60 years respectively[23]. Similarly, human history records many examples of damaging short-term climatic hazards such as storms, floods and droughts. For example, it is not for nothing that the year 1816 was termed 'the year without a summer', for its intense cold was associated with both the Dalton solar minimum and a super-eruption of the Indonesian volcano Tambora in 1815[309]. It is also a matter of record that other somewhat longer events, such as the 1930s Dust Bowl drought in the USA[310], and the severe phases of the 1250-1875 Little Ice Age[301], were marked by extreme hardship and famine. Finally, rapid changes are also recorded from time to time at specific locations by modern instrumental data records. For example, during the 1920s warming in Greenland, at five coastal weather stations the 'average annual temperature rose between 2 and 4°C [and by as much as 6°C in winter] in less than ten years.'[311]

Many of these varied climatic events, whether they are abrupt or manifest themselves as longer-term trends, remain unpredictable – even when viewed with hindsight. Human influence aside, therefore, it is certain that natural climate change will continue in the future, sometimes driven by unforced internal variations in the climate system and at other times forced by factors that we do not yet understand. Such natural changes will in future include both climatic step events, and longer term cooling and warming trends.

Climate change as a natural hazard is therefore as much a geological as it is a meteorological issue. Geological hazards are mostly dealt with by providing civil defence authorities and the public with accurate, evidence-based information regarding events such as earthquakes, volcanic eruptions, tsunamis, storms and floods (which are climatic events), and by adaptation to the effects when an event occurs. The additional risk of longer-term climate change differs from other climate hazards only because it occurs over periods of decades to hundreds or thousands of years. This difference is not one of kind, and neither should be our response plans.

Authorities planning national climate policies therefore need to abandon the alarmist IPCC view of untrammelled global warming, and the illusory goal of preventing it. Instead, real climate change in both directions should be dealt with in the same adaptive way that we treat other natural hazards. Careful planning is needed to identify when a dangerous weather or climate event is imminent (or has started), and ongoing research is needed to foster the development of predictive tools for both sudden and long-term climatic coolings and warmings.

Plan A hasn't worked, won't work, can't work

The global warming issue has become big business indeed for bureaucrats, politicians and business, as well as for scientists and environmental NGOs. It can be estimated that in 2009 Western countries alone were spending at least US$10 billion annually on global warming related research or policy formulation. Such a sum buys a lot of science and influences a lot of adherents. Doug Hoffman and Allen Simmons[312] have reported that in 2009 the United Nations alone was funding 60,000 projects that deal with (human-caused) climate change. And the ascendancy of President Obama to his Washington throne was greeted by a greater than 300 per cent increase in global warming lobbyists to Washington, with 770 companies and interest groups hiring 2,340 lobbyists to influence federal policy on climate change in his first year.

Despite all of this activity and expenditure, the IPCC's Plan A – to restrict future global warming by limiting human carbon dioxide emissions – is dead, for it hasn't worked, it won't work and it can't work

Plan A hasn't worked because it has already been tried at international level, and failed. Remember the Kyoto Protocol? 186 countries have signed up, including Australia in 2008 as tail end Charlie. While it is difficult to establish a summed cost for the expenditures being made by so many disparate countries towards conforming to the protocol, all estimates run to trillions of dollars. And what did the world get for this expenditure? Well, have you looked outside the window lately, and did the climate change? It's scarcely surprising that the answer is no, because all parties to the debate, including the Greens, accepted beforehand that Kyoto was a gesture and not a practical solution. A widely quoted modelling exercise predicted at the time that if all countries meet their Kyoto obligations (which they won't) then an unmeasurable 0.05°C of warming might be prevented by 2050[304]. Some gesture.

Second, we know that Plan A won't work at the level of individual countries, either. For example, Norway was an early mover in introducing carbon dioxide taxation. Norway has had a *de facto* tax of $20-30/tonne on carbon dioxide emissions since the early 1990s, the result of which has been an increase in emissions of 15 per cent[313]. The reason is obvious, and it is that the use of fossil fuels for energy generation and personal transport is inelastic to cost impositions at this level. Individuals and companies alike absorb the extra cost, whingeing awhile, and then shrug their shoulders and get on with life. As economists have pointed out, carbon dioxide will have to be taxed at ten times this rate before a significant diminution of public use will occur. No government could survive such a high cost impost.

Third, Plan A can't work because of the fundamental physical principles that apply to greenhouse gases such as carbon dioxide.

Given the logarithmic relationship that exists between an increase in carbon dioxide and a forced increase in temperature (Fig. 15), doubling the amount of carbon dioxide in the atmosphere will cause only a minor warming of probably less than $1°C$[87]. Consequently, the ambitious plans of, for example, the UK government to cut emissions by 20 per cent by 2020, at the amazing estimated cost of GBP100 billion, are modelled to reduce the temperature in 2100 by $0.0004°C$[314]. Plans of this type are governance run amok, and resemble science fiction rather than prudent public policymaking. In effect, no matter how stringently individual countries reduce their human emissions of carbon dioxide, the effect on future climate will be undetectable.

Clearly, the time has long since passed to take a different approach.

Plan B: a fresh policy approach

It is indeed true that future climate change is an important subject that needs to be approached with appropriate public policy-making. Unfortunately, current policy approaches have been formulated from a combustible combination of poor science, special-interest-group pleading and public hysteria, which together distract from, rather than deal with, the very real risks of natural climate change. Indeed, the risks of natural events and change are almost entirely ignored by the IPCC and by the politicians, press and public who participate in the current climate 'debate'.

A former New Zealand environment minister, Simon Upton, recently wrote[315]:

> It is pointless to apportion blame. But for the sake of environmental credibility and business certainty, the plea has now surely to be that our legislators try to build some constructive middle ground ... Anyone who has studied the [climate change] issue in good faith knows that there are no certainties and that it is a risk management issue.

Mr. Upton is surely right, yet his message is ignored by our current political masters, who continue to pursue the alarmist agenda of global warming extremism even to the point of inflicting yet more pointless damage on an already teetering global economy.

Dealing with future climate hazards, both natural and possibly human-caused, is primarily a matter of risk appraisal – and those risks vary in type and intensity from geographic place to place. Nobody, repeat nobody, lives in a world climate. Indeed a world climate doesn't exist except in the imaginations and computer models of scientists. Instead, all organisms, including humans, live in and are adapted to their local environment and climate. This scientific commonplace has now been understood for more than a hundred years, yet it is all but completely ignored in contemporary public discussions on global warming.

At the start of the twentieth century, Russian geographer Wladimir Koppen used a combination of vegetational zonation and average monthly temperatures and precipitation to develop his now famous map of the 28 climatological zones into which planet Earth can be divided (Fig. 34, inside front cover)[316]. Contemplating this map makes apparent the absurdity of suggesting that a common climate hazard policy should apply to, for example, Indonesia and France, or to Australia and Sweden. Horses for courses, and climate is the same. Also, geographically larger nations, such as Russia, USA and Australia, contain several climatic zones within their own borders, each of which requires a customized climate hazard management plan. Yes, the USA needs a national climate policy, but that policy needs to take account of the fact that the climate hazards are very different in New York, Florida and Arizona, and to provide tailored responses for each. Of course – to avoid setting up new mega-agencies that then, like the IPCC, become powerful lobby groups for their own self interest – climate hazard and response planning may well be better undertaken using existing, appropriately linked and refocused regional or state agencies, rather than of necessity designing giant and costly new bureaucracies at national level.

In essence, every country needs to develop its own understanding of, and plans to cope with, the unique combination of climate hazards that apply to it alone. The idea that there can be a one-size-fits-all solution to deal with future climate change, such as recommended by the IPCC, fails entirely to deal with the real climate and climate-related hazards to which we are all exposed. As Ronald Brunner and Amanda Lynch argue in their new book[317], we need to use adaptive governance to produce response programs that cope with hazardous climate events as they happen, and that encourage diversity and innovation in the search for solutions. In such a fashion, the highly contentious 'global warming' problem can be recast into an issue in which every culture and community around the world has an inherent interest.

An example of the pragmatic approach that is required is provided by a recent study of the risk of sea-level rise in eastern USA, led by EPA researcher Jim Titus[318]. Titus's research team designated all coastal plain land within one metre of sea-level into four categories: developed, intermediate (likely to be developed), undeveloped and protected from development. The first two categories, which comprise 60 per cent of the coastal land area between Florida and Massachusetts, were judged likely to be protected from future sea-level rise by human intervention, which leaves the remaining 40 per cent of low-lying land vulnerable to flooding. Noting that only about a quarter of this area is currently conservation protected, Titus *et al.* recommended that as much as possible of the remaining three-quarters should either be protected by conservation, or else left alone to respond naturally by the inland migration of the relevant ecosystems.

The main point implicit in Titus *et al.*'s paper is the importance of planning land-use well ahead, and in a way that recognizes both the inevitability of natural change and that attempting to stop sea-level rise (or, in other contexts, climate change) is simply not a practical option. Natural climate changes are without doubt going to continue to affect our planet, and from time to time these changes will wreak human and environmental damage. Future temperature changes will include cooling trends, warming trends and sudden step events. Extreme weather events and their consequences, and prolonged inconveniences such as droughts, are natural disasters of similar character to earthquakes, tsunamis and volcanic eruptions, in that in our present state of knowledge they can neither be predicted far ahead nor prevented once they are under way.

Interestingly, commentators from both main sides of the political spectrum are finally embracing the need for Plan B climate policies that are based upon adaptation. Thus, writing in the *Wall Street Journal* in late 2009[319], former chancellor Lord Nigel Lawson commented that:

> The time has come to abandon the Kyoto-style folly that reached its apotheosis in Copenhagen last week, and move to plan B. And the outlines of a credible plan B are clear. First and foremost, we must do what mankind has always done, and adapt to whatever changes in temperature may in the future arise.

And surprisingly, given its long-term editorial attitudes, *Nature* agrees with him, commenting[320] about three weeks later that:

> Science could help untangle this politically impossible dilemma by moving away from its obsession with predicting the long-term future of the climate to focus instead on the many opportunities for reducing present vulnerabilities to a broad range of today's — and tomorrow's — climate impacts. Such a change in focus would promise benefits to society in the short term and thus help transform climate politics.

National natural hazard agencies

The existence of such natural hazards as those just discussed is the prime reason why civil defence agencies exist. Throughout the world, such agencies consist of a mix of national and regional organizations and volunteer groups, of which I will briefly discuss just three.

In USA, national emergency planning is the responsibility of the Federal Emergency Management Agency (FEMA), which has as its primary mission:

> To reduce the loss of life and property and protect the Nation from all hazards, including natural disasters, acts of terrorism, and other man-made disasters, by leading and supporting the Nation in a risk-based, comprehensive emergency management system of preparedness, protection, response, recovery, and mitigation.

One's enthusiasm for developing other national equivalents to FEMA is somewhat tempered by the rather illogical inclusion of anti-terrorism in its list of duties, and by the strong criticism that FEMA received for its inadequate response to the Hurricane Katrina disaster. However, remove anti-terrorism from FEMA's brief and you are left with a statement that would serve as a fair model for the development of other national natural disaster emergency response agencies – the key point being the need for preparedness, protection, response, recovery and mitigation (in its true, not IPCC, meaning) to all natural hazards.

In contrast, other countries such as Australia lack a strong national emergency agency. Rather, the national Emergency Management Australia (EMA), which for some strange reason is administered through the federal attorney-general's department, has as its mission: 'Provide national leadership in the development of emergency management measures to reduce the risk to communities and manage the consequences of disasters,' but this is in the restricted context of EMA being a training rather than an implementation agency. On the ground, natural hazards, such as floods and bushfires, are dealt with using a complex mix of federal and state government and volunteer groups. Though not unhealthy of necessity, such complex overlapping of hazard responsibilities can cause organizational turf wars, and leads to overlaps or gaps in emergency response to particular disasters and confusion in the chain of command; it also tends to be financially inefficient.

One of the world's best-practice organizations for monitoring emergency civil defence lies with Australia's near neighbour, New Zealand, which has established a widely admired GeoNet organization to advise on and manage environmental hazards in a country which has many[321]. GeoNet maintains a network of geophysical monitoring stations countrywide, and provides civil authorities and the public with accurate, evidence-based information about hazards like earthquakes, volcanic eruptions, tsunamis, landslides and floods. Though longer term climate change has so far not been included in GeoNet planning, it differs from the hazards that are covered only in the extended decadal time-scales over which a deleterious trend might occur. GeoNet already deals with short-term weather events such as storms and floods, and it could easily and cost-effectively manage the risks of longer-term climatic changes.

As for other natural planetary hazards, then, policies to cope with climate change should be based upon adaptation to the change as it happens, including appropriate mitigation of undesirable socio-economic effects. The appropriate public policy response to climate hazard is, first, to monitor climate change accurately in an ongoing way; and, second, to respond and adapt to any changes – including short-term events and long-term warmings and coolings – in the same way that we deal with other hazardous natural events such as earthquakes and volcanic eruptions.

Conclusion

To focus on the chimera of human-caused greenhouse warming while ignoring the real threats posed by the natural variability of the climate system is self-delusory. Instead, the realities that global climate is currently cooling, that it will both warm and cool again in the future, and that unpredictable, unpreventable and damaging 'weather' events will continue to recur, need to be recognized.

It is therefore time to move away from stale 'he-says-she-says' arguments about whether human carbon dioxide emissions are causing dangerous warming, and on to designing effective policies of hazard management for all climate change, based on adaptation responses that are tailored for individual countries or regions. The key issue on which all scientists agree is that natural climate change is real, and every year brings new examples that exemplify the substantial human and environmental damage that it can cause. By their very nature, strategies that can cope with the dangers and vagaries of natural climate change will readily cope with human-caused change too should it ever become manifest.

Even were generous funding to be provided for the implementation of national hazard warning and disaster relief schemes, the overall costs would be orders of magnitude less than those caused by the introduction of unnecessary and ineffectual emissions trading schemes and other 'anti-carbon dioxide' measures. To boot, contingent damage to the world economy, the standard of living and the world food supply would be avoided.

12 Who are the climate denialists now?

For more than a decade, we've been told that there is a scientific 'consensus' that humans are causing [dangerous] global warming, that 'the debate is over' and all 'legitimate' scientists acknowledge the truth of global warming. Now we know what this 'consensus' really means. What it means is: the fix is in.

This is an enormous case of organized scientific fraud, but it is not just scientific fraud. It is also a criminal act. Suborned by billions of taxpayer dollars devoted to climate research, dozens of prominent scientists have established a criminal racket in which they seek government money – Phil Jones has raked in a total of £13.7 million in grants from the British government – which they then use to falsify data and defraud the taxpayers. It's the most insidious kind of fraud: a fraud in which the culprits are lauded as public heroes....

The damage here goes far beyond the loss of a few billions of taxpayer dollars on bogus scientific research. The real cost of this fraud is the trillions of dollars of wealth that will be destroyed if a fraudulent theory is used to justify legislation that starves the global economy of its cheapest and most abundant sources of energy.

This is the scandal of the century. It needs to be thoroughly investigated – and the culprits need to be brought to justice.

(Robert Tracinski, 2009)[322]

As the preparation of this book was in its final stages, two epochal, climate policy-related events were underway. The first was the Climategate scandal, a term that refers to the leaking of a package of emails and ancillary documents from the Hadley Climatic Research Unit (CRU) at the University of East Anglia, the research group that supplies the IPCC with their main global temperature record. The second was the convening in Copenhagen in December 2009, of the IPCC's COP-15 climate conference, a conference that was intended to deliver to the world a carbon dioxide-limiting successor to the Kyoto Protocol.

The state of public opinion prior to Climategate

Prior to the swing-year of 2009, many responsible and concerned people in our society believed the following three propositions to be true:

> Late twentieth century human-caused global warming is an established fact that demands an urgent political solution;

> Carbon dioxide is a harmful pollutant of the atmosphere, and the primary cause of the perceived warming;

> An overwhelming consensus of qualified scientists support the view that dangerous, contemporary, human-caused climate change is occurring.

Indeed, how could people have believed otherwise when they were constantly bombarded with messages to this effect in print, on radio, on television, at the movies, by their own governments and even by the United Nations? Huge amounts of money have been spent, by proselytising environmental groups and climate-related scientific and commercial interests, to command the daily media with the message that 'global warming is here, it's bad, and it's YOUR fault'.

Accompanying this media blitz has been the allied politicisation of scientists, science bureaucrats, science organizations and science journals, as described in Chapter 7. Under a remorseless worldwide pressure that government funding should be supplied to scientists mainly for research on 'real societal problems', scientists and their organizations now compete to be world class not at science, nor even at problem solution, but rather at problem generation – with a special premium attached to scary problems that can be linked to atavistic guilt in the general citizenry. The global warming threat has thereby been turned into the mother of all environmental scares, on which more than US$10 billion annually is now being expended for 'research', worldwide.

You don't have to be a cynic to observe that with such political pressure and large sums of money at stake it is going to be a brave scientist who stands up to say that human-caused global warming isn't a problem, let alone to venture the thought that warming might well be beneficial if it ever actually occurs. Nonetheless, and despite the blitzkrieg of abuse waged against them, by 2009 persistent critical analysis by independent scientists, aided by the power of internet communication, had stripped the global warming emperor naked and in the process shown that the three assertions listed at the head of this section are false. Climategate is now proving to be the final Act in the theatre of science scepticism that has dogged the IPCC ever since its creation in 1988.

Climategate

On 12 October 2009, the climate correspondent for BBC's *Look North* in Yorkshire and Lincolnshire, Paul Holmes, received an unusual email[323]. In response to a 9 October article that he had written entitled *What has happened to global warming?*[324], Holmes appears to have been the first reporter to receive the explosive package of hacked or leaked emails from CRU at the University of East Anglia, around which the Climategate crisis subsequently erupted[325]. Perhaps already under pressure for having broken the alarmist BBC corporate line with his earlier article, Holmes chose not to write the scoop that begged, but instead 'passed the news on to some of my colleagues in the BBC's environment and science team, including our environment analyst Roger Harrabin.' It was over a month later before the BBC made its first comment on the email package, on its news website on 20 November, and even then it misreported the matter as a hacker's attack and revealed no information as to the content of the messages[326]. Mr. Harrabin himself did not cover[327] the issue until 24 November. Notably, therefore, the BBC's coverage of the story post-dated the first public comments about the emails, which were made on 19 November by the authors of three US climate websites[328], *The Air Vent, Watts Up With That?* and *Climate Audit.* Thus the BBC's coverage of Climategate was entirely reactive to stories that first appeared elsewhere.

To understand this lack of interest in a news story of such magnitude requires a rewind of the tape to a famous private meeting that was organized by Harrabin on 26 January 2006 at the BBC Television Centre[329]. Titled *Climate Change – the Challenge to Broadcasting*, the meeting was co-hosted by the director of television (Jana Bennett) and the director of news (Helen Boaden), and held under the auspices of the BBC and two environmental lobby groups – The International Broadcasting Trust and the Cambridge Media and Environment Programme. Former Royal Society President, Robert May, was the keynote speaker and delivered a briefing on global warming to an audience that included about thirty key BBC staff and executives and about thirty invited guests, most of whom were environmental activists. The (untrue) message delivered was that the science supporting global warming alarm was so certain that it was the BBC's public duty to cease providing airtime to alternative viewpoints.

One participant at the meeting recalled: 'I found the seminar frankly shocking. The BBC crew (senior executives from every branch of the Corporation) were matched by an equal number of specialists, almost all (and maybe all) of whom could be said to come from the "we must support Kyoto" school of climate activists,' continuing, 'I was frankly appalled by the level of ignorance of the issue which the BBC people showed.' Not only did the BBC subsequently fall into line with the preposterous idea put by Lord May, but, as the biased news coverage of global warming by public broadcasters ever since has shown, the same attitude was soon propagated to national broadcasters in at least New Zealand, Australia and Canada via the invisible professional networks that link commonwealth broadcasters and Royal Societies worldwide.

What, then, did the leaked package of CRU papers contain, and why was their release so significant? The papers comprised more than 60 Mb of emails, reports and computer code relating to CRU's research activities. The package is now posted on several server sites[330], it has its own nominate website, and other extensive analyses[331] have also been undertaken of its content and implications – about which the first book has already been written[332], with others in train. The most disturbing material in the package, however, is not the emails around which most of the sensation has been created, but the computer code. This code, and accompanying programmers' notes, shows unequivocal evidence for data manipulation aimed at suppressing the mediaeval warming pattern, as required by IPCC orthodoxy, in place of allowing the intrinsic temperature pattern to emerge. But rather than my picking and choosing from the parts of the dossier that I have chanced to read, let me instead quote the considered opinion of a respected investigative journalist from New Zealand who has systematically read every one of the emails[333].

> Having now read all the Climategate emails, I can conclusively say they demonstrate a level of scientific chicanery of the most appalling kind that deserves the widest possible public exposure.

The emails reveal that the entire global warming debate and the IPCC process is controlled by a small cabal of climate specialists in England and North America. This cabal, who call themselves 'the Team,' bully and smear any critics. They control the 'peer review' process for research in the field and use their power to prevent contrary research being published.

The Team's members are the heart of the IPCC process, many of them the lead authors of its reports.

They falsely claim there is a scientific 'consensus' that the 'science is settled,' by getting lists of scientists to sign petitions claiming there is such a consensus. They have fought for years to conceal the actual shonky data they have used to wrongly claim there has been unprecedented global warming this past 50 years. Their emailed discussions among each other show they have concocted their data by matching analyses of tree rings from around AD 1000 to 1960, then actual temperatures from 1960 to make it look temperatures have shot up alarmingly since then, after the tree rings from 1960 on inconveniently failed to match observed temperatures.

It is clear that the game is up, and not just for the CRU and its BBC supporters. Surprisingly, perhaps, the first domino that fell was in the US, where on 28 November the University of Pennsylvania[334] announced that an investigation would be held into the research activities of Michael Mann, the notorious senior author of the discredited hockey-stick temperature graph and a participant in the CRU email circle.

Shortly afterwards, a second domino dropped at the CRU itself, where the director, Professor Phil Jones, stood aside from his position on 1 December, pending the result of another independent enquiry[335].

On the very same day, domino three fell on the other side of the world, where, just one week before the opening of the

Copenhagen climate conference, the parliamentary opposition in Australia elected a new leader, Tony Abbott, who was committed to voting against a hotly disputed emissions trading bill – the first such 'climate sceptic' leader of a mainstream Western political party in recent time[336].

Domino four, on 4 December, was a statement by the chair of the IPCC, Rajendra Pachauri, saying that he also favoured an enquiry into the CRU emails, noting, 'We certainly don't want to brush anything under the carpet. This is a serious issue and we will look into it in detail.'[337]

Fifth domino in line was the British Meteorological Office, which announced on 5 December that it would undertake a new analysis of the last 160 years of surface temperature data, because public confidence in the science behind man-made global warming has been shattered by the leaked e-mails[338].

Domino six takes us back to the BBC, whose governing trust announced on 6 January a review of the impartiality of BBC's science coverage, including such controversial topics as genetic crop modification and climate science[339].

Domino seven saw the broadcast on US TV network KUSI on 14 January of a one-hour long Special Report on Global Warming by John Coleman, the doyen of weather broadcasters, in which it was shown that Climategate-like data tampering extends also to the US temperature databases[340].

An eighth domino fell into place on 17 January with another revelation of IPCC incompetence, this time regarding untrue statements about Himalayan glaciers melting contained in their 2007 report[341], which caused Pachauri to promise yet another investigation into what had gone wrong accompanied by the self-delusory statement that 'We have to see that its [the IPCC's] gold-plated standard is maintained.'[342]

Domino nine was the announcement on 22 January that the UK House of Commons Science and Technology Select Committee was to hold a formal enquiry into the Climategate affair[343] to investigate the adequacy of the terms of reference of the

University of East Anglia review, the implications of the matter for the integrity of scientific research generally, and to judge how independent other international estimates of global temperature are from those of CRU.

Stopping at ten dominos is convenient for this book, though it is surely only a temporary number, and number ten duly fell on 4 February, when the Indian environment minister, Jairam Ramesh, announced[344] that his government would no longer rely solely on IPCC advice about climate change, but would instead use advice to be provided by a new Indian Network on Comprehensive Climate Change Assessment (INCCCA) that would draw input from 125 research groups throughout India; with a resounding thud, the first nail was thus driven into the IPCC's coffin.

Whilst these dominoes were toppling, and doubtless partly driving their fall, news of the CRU email release was spreading rapidly via the internet, with Google listing more than 20 million page hits for 'climategate' by early December. In marked contrast, most mainstream media sources (including the BBC) ran the story only slowly, and reluctantly. Later in December and in January, however, there was a marked increase in articles and programmes attempting to make balanced comment on either Climategate, or on global warming alarmism more generally. In the meantime, evidence of suspicious warmth-inducing corrections to local temperature records had emerged also in Australia[345], New Zealand[346] and USA-Canada[60], accompanied by calls for a congressional investigation into the National Oceanic and Atmosphere Administration (NOAA) and the National Aeronautics and Space Administration (NASA)[347], the two government agencies responsible for furnishing US national climate data.

A new pack of dominoes contains 28 tiles, and more than one pack is clearly going to be needed to represent all the consequences that will eventually flow from the Climategate affair. Interested readers should check the websites listed in the recommended references (p. 256), where future developments will be described and discussed.

Meanwhile, as big a problem as the attitudes revealed among climate researchers by the Climategate emails is the attitudes that are endemic at premier scientific journals like *Nature* and *Science*, as conveyed in their editorials and articles about the leak. Exhibiting stunning ignorance, a *Nature* writer commented on 20 January 2010, on the 'the extreme rate of the twentieth-century temperature changes' thereby in one phrase denying the worth of the contributions of the literally thousands of skilled scientists whose findings underpin Chapter 1 of this book. The editorial in the same edition is scarcely less breathtaking in its assertion that '... there is little uncertainty about the overall conclusions: greenhouse-gas emissions are rising sharply, they are very likely to be the cause of recent global warming and precipitation changes, and the world is on a trajectory that will shoot far past 2°C of warming unless emissions are cut substantially,' a cracked if not entirely fragmented old recording that is flatly contradicted by the copious evidence discussed in the first half of this book, especially Chapters 2 and 3. Cleaning up the climate act is going to remain difficult until those those managing our leading scientific journals and scientific societies come to grips with the empirical reality of the world as it is, rather than doggedly pursuing their particular, beloved virtual reality of how it might one day be.

Copenhagen

Unlike Climategate (which turned out to trump it), the Copenhagen COP-15 climate conference held on 7-18 December was anticipated long in advance to be *the* climate event of 2009. Endless pre-conference tasters, teasers, preludes, opinion pieces, feature articles and editorials, not to mention advertisements, resulted in worldwide saturation news coverage well before more than 5,000 journalists descended upon Copenhagen to cover the event itself. The push had been on for years, and it was for an international carbon dioxide trading (aka taxation) scheme to follow on from Kyoto.

Meaningful news copy of progress towards that cause was, however, hard to find. It had been obvious long before the conference started that an impasse would develop between the interests of the irrational Western environmental lobby groups and the intransigently pragmatic viewpoints of China, India and Brazil – who provided leadership for the interests of all developing countries; and so it proved. Instead of smooth progress, the press reported inadequate facilities, over-zealous security, street scuffles, angry negotiating standoffs and a lack of any meaningful compromise solutions. This state of affairs did not improve when more than 100 heads of state flew in to participate in the second week of the conference, the highlight of which appears to have been an anti-capitalist diatribe delivered by Venezuelan President Hugo Chavez to a standing ovation.

The eventual outcome of the conference, hastily put together by the US, China, India, Brazil and South Africa (note the absence of Europe) as other more ambitious negotiations foundered, became an anodyne statement of aims. Namely, to cut carbon dioxide emissions so as to limit any future rise in global temperature to 2°C (there being no convincing scientific basis for either part of that proposition), and to provide US$100 billion a year to support the anti-global warming efforts of poor nations by 2020 (without any indication as to who would pay what, or why). After the hype that had preceded the conference, to term this agreement 'a meaningful and unprecedented breakthrough,' as President Obama did at his press conference, was simply casuistry.

More realistically, the London *Financial Times*[348] termed the Copenhagen outcome 'dismal', 'a fiasco', 'worse than useless' and 'the emptiest deal one could imagine, short of a fist fight', adding that 'If you draw the world's attention to an event of this kind, you have to deliver, otherwise the political impetus is lost. To declare what everybody knows to be a failure a success is feeble, and makes matters worse.' And on the other side of the Atlantic the *Wall Street Journal* commented[349] that the global effort to combat climate change remained stuck in essentially the same place, and that 'far from resolving the issue, the Copenhagen conference set up months more of international haggling over what to do about climate change.'

Perhaps the final nail in Copenhagen's coffin was driven in by India and China, when they announced on 23 January that they wouldn't sign it[350]. But the last word on the COP-15 conference should be left to British climate commentator Benny Peiser[351], because no one has summarized the significance of the event more clearly than he:

> The failure of the UN climate summit in Copenhagen is a historical watershed that marks the beginning of the end of climate hysteria. Not only does it epitomise the failure of the EU's environmental policy, it also symbolises the loss of Western dominance. The failure of the climate summit was not only predictable – it was inevitable.
>
> There is no way out from the cul-de-sac into which the international community has manoeuvred itself. The global deadlock simply reflects the contrasting, and in the final analysis irreconcilable, interests of the West and the rest of the world. The result is likely to be an indefinite moratorium on the international climate legislation. After Copenhagen, the chances for a binding successor of the Kyoto Protocol are as good as zero.
>
> The Asian-American Accord connotes a categorical No to legally binding emission targets. This means that a concrete timescale for the curtailing of global CO_2 emissions, not to mention the reduction of the CO_2 emissions, has been kicked into the long grass. The green dream of industrial de-carbonisation has been postponed indefinitely.

Who are the denialists now?

As I write this, in January 2010, the political, bureaucratic and scientific leaders of Western European and other industrialized nations have yet to fully absorb the outcomes of Climategate and Copenhagen. Together with the managers of many scientific journals and societies, they are in disarray, deep retreat and denial.

They deny that the Earth's climate is cooling; they deny that the climate models on which their global warming policies are based are worthless as predictive tools; they deny that the IPCC and its advice are flawed beyond repair; they deny that the Copenhagen Conference was a failure; they deny that carbon dioxide is an environmental benefice; they deny that Climategate is any more than an isolated, minor squabble among a few climate research cognoscenti; they deny that they have allowed their young people to be educationally brainwashed about global warming; they deny that the science research community has been corrupted by their agenda-driven funding requirements; they deny that government science-related organizations, at their behest, have been acting as propagandists for eco-evangelistic causes; they deny that windfarms and solar power are environmentally damaging and uneconomic for baseload power generation; they deny that nuclear power is the 'greenest' baseload energy source available; they continue to strive to deny public voice to independent scientific viewpoints on climate change; and, above all, they deny that they are wrong in their continued assertions that human-caused global warming is an identified and deadly danger.

Of course they do, because the loss of face that will occur otherwise is simply beyond their contemplation. Nevertheless, and squirm though they will, our current generation of political leaders is going to lose the false battle of global warming in which they have foolishly engaged. For, inch by inch, public opinion is finally turning, and recent polls from across the Western world show that a majority of voters no longer buy the global warming misinformation that they have been fed for so long. Slowly, the realization is dawning that the global warming scare has been a colossal scientific and political scam.

Only dimly do OECD politicians yet perceive that their arrogant post-colonial belief that they would save the world from climate disaster has delivered to developing countries a rod of climate guilt that will long be used to belabour Western backs. Compounding the damage is that the same political leaders have also betrayed their own youth, by allowing – indeed, in many cases encouraging – them to be influenced by propaganda on this and other environmental issues through failing education systems and degraded state science bureaucracies and public agencies.

The loss of political face and formerly 'smart' investment money that is now impending will be so great that it is difficult to predict exactly how, or how fast, the global warming castle will tumble, but collapse it will. Perhaps the best way of encouraging the bricks to fall, and at the same time to do some good for the future, would be to accept earlier advice[352] (and that of Robert Tracinski[322]) to institute judicial commissions of enquiry into the state of climate science and climate policy, if not science and policy more generally, in individual Western countries.

Conclusion

The late 2009 events of Climategate and Copenhagen have laid bare the corruption of the evangelist global warming cause, torpedoed the chances of international agreement on the control of carbon dioxide emissions, and rewritten the map of climate change policy. It might be said, therefore, that the need for a book such as this one, which attempts to explain both the basic science and the socio-political pathology that surrounded the 1988-2009 global warming scare, has now passed.

Nothing could be further from the truth. For the fact that dangerous human-caused global warming is being perceived less and less as an imminent hazard in no way reduces the very real threats and dangers posed by future natural climate events, nor from the need to possess a contingent policy to deal with significant human-caused climate change should it, after all, chance to emerge. Accordingly, far from being rendered redundant by Climategate and Copenhagen, the policy actions recommended in Chapter 11 of this book have actually been given sharpened relevance.

Final Coda

'*Well, Grandpa, was the great global warming scare a conspiracy, then,*' I am asked, '*and what should Mummy do now with the annual donation that she formerly gave to our favourite environmental organization?*'

I have been asked the first question many times before, and it has always made me pause. Surely, I would think, it is simply silly to suggest that so many, so well-educated persons in so many countries could all have been so ignorant of basic science principles, and therefore so easily conned by a so obviously alarmist climate credo – lavishly funded, skilful, pervasive and media-supported though the global warming propaganda campaign has been.

But time, Climategate and Copenhagen have all passed by, and now I nod, slowly but unhesitatingly, in response to the question.

'*Yes,*' I answer, '*it is the greatest self-organized scientific and political conspiracy that the world has ever seen, made worse by the fact that many of the people taken in by it had only the best of intentions, but lacked the science education to see through the scam.*'

'*And Mummy should ensure that any charitable donations that the family makes in future go toward providing the underprivileged people of the world – there are more than 1,500,000,000 of them – with clean drinking water, adequate sanitation, basic education and basic health care. For lifting the poor out of their poverty, and helping them to generate wealth for themselves, is the only sure way to protect Earth's future environment.*'

13 Postscriptum

A balanced statement regarding climate change

The consensus view of experienced scientists is undoubtedly that both the extreme 'alarmist' and the extreme 'denialist' views of human-caused global warming are incorrect. The majority of scientists, most of whom are independent of the IPCC, espouse views that represent a balanced summary of the available scientific information. Such a view neither inflates nor underestimates the risk of human-caused change, and at the same time takes proper account of the demonstrably severe risks of natural climate change.

The following, and final, quotation can be taken as a case in point. It was written by a group of authors whose acknowledged expertise includes climatology, atmospheric physics and geology, as befits a topic that ranges across such a great range of scientific subdisciplines[353]. Very similar views have been espoused by the groups of 103 and 143 professional persons who have written to the Secretary General of the United Nations about climate policy (Letters of November 2007 and December 2009), by the 197 expert persons who signed the Manhatten Declaration, by the 650 scientists listed in a 2009 US Senate report (Morano, 2009), and by many other scientists both on and off the public record[354].

> Climate changes naturally all the time. Human activities have an effect on the local climate, for example in the vicinity of cities (warming) or near large areas of changed land usage (warming or cooling, depending upon the changed albedo). Logically, therefore, humans must have an effect on global climate also. This notwithstanding, a

distinct human signal has not yet been identified within the variations of the natural climate system, to the degree that we cannot even be certain whether the global human signal is one of warming or cooling. Though it is true that many scientists anticipate that human warming is the more likely, no strong evidence exists that any such warming would be dangerous.

The gentle global warming that probably occurred in the late twentieth century falls within previous natural rates and magnitudes of warming and cooling, and is prima facie quite unalarming, especially when consideration is given to the likelihood that the historic ground temperature records used to delineate the warming are warm-biased by the urban heat island and other effects. Once corrected for non-greenhouse climate agents such as El Nino and volcanic eruptions, the radiosonde (since 1958) and satellite (since 1979) records show little if any recent warming and certainly none of untoward magnitude.

Atmospheric carbon dioxide is indeed a greenhouse gas, but the empirical evidence shows that the warming effect of its increase at the rates of modern industrial emission and accumulation is minor, given an assumed pre-industrial level of about 280 ppm and noting the established logarithmic relationship between gas concentration increases and warming. As one such empirical test, it can be noted too that no global increase in temperature has now occurred since 1998 despite an increase in carbon dioxide concentration over the same eight years of about 15 ppm (5 per cent).

Putative human influence aside, it is certain that natural climate change will continue, sometimes driven by unforced internal variations in the climate system and at other times forced by factors that we do not yet understand. The appropriate public policy response is, first, to monitor climate accurately in an ongoing way; and, second, to respond and adapt to any changes – both warmings and the likely more damaging coolings – in the same way that we cope with other natural events such as droughts, cyclones, earthquakes and volcanic eruptions.

In dealing with the certainties and uncertainties of climate change, the key issue is prudence. The main certainty is that natural climate change will continue, and that some of its likely manifestations – sea-level rise and coastal change in particular locations, for example – will be expensive to adapt to. But adapt we must and will. Moreover reducing vulnerability to today's climate-sensitive problems will also help the world cope with future challenges from climate change whether that is due to natural variability, anthropogenic greenhouse gas emissions or other human causes. The most prudent way of ensuring that happens is to build wealth into the world economy and to be receptive to new technologies. This will not be achieved by irrational restructuring of the world's energy economy in pursuit of the chimera of 'stopping' an alleged dangerous human-caused climate change that, in reality, can neither be demonstrated nor measured at this time.

The world has the IPCC, for at least a little longer. It also has innumerable national greenhouse offices, ministries of climate change, state greenhouse offices, specialist climate change sections within government departments, bureaus of meteorology, national and international science organizations with climate alarmist views, and an untold number of other climate change research groups, organizations and lobbyists. What it does not seem to have is measurable human-caused climate change.

Present public policy on global warming remains about where the science was at in 1990 – looking for, and reacting to, ghosts. It is time to implement Plan B – preparation for and adaptation to all climate events and change as they occur, irrespective of their cause.

Acknowledgements

Thank you to my early academic mentors, stratigrapher Doug Campbell, petrologist Doug Coombs, structural geologist Win Means and palaeontologist and historian of science Martin Rudwick, who many decades ago set out to teach me to think clearly, and above all honestly, about science.

Climate science is a demanding new discipline-complex that spans more than one hundred specialties, of which most scientists are expert in but one or two. Thank you, therefore, to the large group of my climate rationalist ('sceptical') colleagues from other scientific disciplines, worldwide; via email, exchange of papers, web communication and interaction at conferences they have patiently taught me the rudiments of many of the aspects of climate science which lie outside my immediate expertise. It is a privilege to have been given this education, and to be received as a member of that body of persons who believe that public policy should be based upon the verifiable facts and relationships of nature, rather than upon contrived scientific or political spin.

A particular thank you, too, to the four close colleagues with whom I helped advise Senator Stephen Fielding during the 2009 emissions trading debate in the Australian parliament, Bill Kininmonth, David Evans, Stewart Franks and Willie Soon. Other scientists to whom I owe a particular debt, and thank for their help over many years, include Jennifer Marohasy, Walter Starck and Peter Ridd. Two distinguished policy organizations, the Institute for Public Affairs (Melbourne) and the Heartland Institute (Chicago), have also striven mightily to further an informed public discussion of the global warming issue, and I thank their staff for the important and unflagging support that they continue to provide to many independent-minded scientists.

Several of the chapters in this book have their origins in invited public essays on climate change, especially one requested by the Royal Society of New Zealand in 2006 and others published in *Quadrant, The Australian*, the UK *Sunday Telegraph* and the journal of the Economic Society of Australia (Queensland), *Economic Analysis and Policy*. I thank the editors who solicited the original pieces for their encouragement and support, and Tom Stacey, Clemmie Jackson-Stops and Charles Powell of Stacey International Press for their skilled help during the editing and publication processes. I also thank most warmly John McLean, Tom Segalstad, Willie Soon and Raphael Wust for their helpful and timely critical comments on parts of the draft manuscript of this book.

Finally, and most important of all, thank you to my wife Anne and to a small number of close friends and colleagues for providing the kind of nurturing support that allows lucky authors to focus single-mindedly, some would say obsessively, on the task at hand.

Bob Carter
Townsville, January 2010

Notes on the figures

1. After: Mix, A.C., Pisias, N.G., Rugh, W., Wilson, J., Morey, A. & Hagelberg, T., 1995. Benthic foraminiferal stable isotope record from Site 849, 0-5 Ma: Local and global climate changes. In: Pisias, N.G., Mayer, L., Janecek, T., Palmer-Julson, A. & van Andel, T.H. (eds.), *Proc. ODP, Scientific Results* 138, College Station, TX (Ocean Drilling Program), 371-412.

    Mix, A.C., Le, J. & Shackleton, N.J., 1995. Benthic foraminifer stable isotope stratigraphy of Site 846: 0-1.8 Ma. In: Pisias, N.G., Mayer, L., Janecek, T., Palmer-Julson, A. & van Andel, T.H. (eds.), Proc. ODP, *Scientific Results* 138, College Station, TX (Ocean Drilling Program), 839-56.

2. After: Salamatin A.N., Lipenkov V.Ya., Barkov N.I., Jouzel J., Petit J.R., Raynaud D., 1998. Ice-core age dating and palaeothermometer calibration based on isotope and temperature profiles from deep boreholes at Vostok Station (East Antarctica). *Journal of Geophysical Research* 103, N D8, 8963-77.

3. Redrawn after an original figure in *New Scientist*, June 17, 1989.

4. After: Morgan, V.I., 1985. An oxygen isotope - climate record from the Law Dome, Antarctica. *Climate Change* 7, 415-26.

    Stuiver, M., Grootes, P.M. & Braziunas, T.F., 1995. The GISP2 ^{18}O climate record of the past 16,500 years and the role of the sun, ocean and volcanoes. *Quaternary Research* 44, 341-54.

5. Illarianov, A. (Dec., 2004), after Alley, R.B., 2004. GISP2 Ice Core Temperature and Accumulation Data, NOAA.

6. After: Loehle, C. & McCulloch, J.H., 2008. Correction to: A 2000-Year Global Temperature Reconstruction Based on Non-Tree Ring Proxies. *Energy and Environment* 19, 93-100.

7. Illarianov, A. (Dec., 2004), after Alley, R.B., 2004. GISP2 Ice Core Temperature and Accumulation Data, NOAA.

8a: After: NOAA National Geophysical Data Center (NGDC), 2010. Solar Data Services. Sunspot Numbers. *http://www.ngdc.noaa.gov/stp/ SOLAR/ftpsunspotnumber.html*.

8b.   After: Woods, T.N. & Lean, J. 2007 (Oct. 30). Anticipating the next decade of Sun-Earth system variations. *EOS (Transactions AGU) 88 (44), doi:10.1029/2007EO440001.*

9.   After: Neff, U., et al., 2001. Strong coherence between solar variability and the monsoon in Oman between 9 and 6 kyr ago. *Nature 411, 290-3.*

10.   After: Carbon Dioxide Information Analysis Center (CDIAC), 2010. *http://cdiac.ornl.gov/ftp/ndp030/global.1751_2006.ems.*

        Intergovernmental Panel on Climate Change (IPCC) 2001. *Climate Change 2001: The Scientific Basis. Contribution of Working Group I to the Third Assessment Report of the IPCC.* Ed.: J.T. Houghton, *et al.*, Cambridge University Press. Synthesis Report, Fig. 2.3.

11a.   After: Thorne, P.W., *et al.*, 2005. Revisiting radiosonde upper air temperatures from 1958 to 2002. *Journal of Geophysical Research. 110: D18105, doi:10.1029/2004JD005753.*

11b.   Satellite MSU temperature data after Spencer, R., UAHuntsville, Atmospheric Science Department; *http://vortex.nsstc.uah.edu/data/msu/t2lt/uahncdc.lt.*

12.   Satellite MSU temperature data after Spencer, R., UAHuntsville, Atmospheric Science Department; *http://vortex.nsstc.uah.edu/data/msu/t2lt/uahncdc.lt.* Carbon dioxide data after Carbon Dioxide Information Analysis Center (CDIAC); *http://cdiac.ornl.gov/ftp/ndp030/global.1751_2006.ems.*

13.   After: Davis, J.C. & Bohling, G.C., 2001, The search for pattern in ice-core temperature curves. *AAPG Studies in Geology* 47, 213-29.

14.   After: Royer, D.L., *et al.*, 2004. CO_2 as a primary driver of Phanerozoic climate. *GSA Today* 14 (3), 4-10. Fig. 2.

        Berner, R.A. & Kothavala, Z., 2001. GEOCARB III: A revised model of atmospheric CO_2 over Phanerozoic time. *American Journal of Science* 301, 182–204.

15.   After: Archibald, D., 2008 (March). Solar Cycle 24: Implications for the United States. *2nd Heartland Conference on Climate Change,* New York, p. 22.

16.   After: Roy Spencer, 2008 (Dec. 28). Global Warming as a Natural Response to Cloud Changes Associated with the Pacific Decadal Oscillation (PDO). *http://www.drroyspencer.com/research-articles/global-warming-as-a-natural-response/.*

17. After: Soon, W., 2007. Implications of the secondary role of carbon dioxide and methane forcing in climate change: past present and future. *Physical Geography* 28, 97-125, Fig. 1.

18. Based on the IPCC's 4th Assessment Report, and information and a figure at *http://homepage.mac.com/uriarte/carbon13.html*.

19. After: Camoin, G.F., Montaggioni, L.F. & Braithwaite, C.J.R., 2004. Late glacial to post glacial sea levels in the Western Indian Ocean. *Marine Geology* 206, 119-46.

20. After: Intergovernmental Panel on Climate Change (IPCC) 2001. *Climate Change 2001: The Scientific Basis. Contribution of Working Group I to the Third Assessment Report of the IPCC.* Ed.: J.T. Houghton, *et al.*, Cambridge University Press. Chapter 11, Fig. 11.7.

21. After: Bureau of Meteorology (Australia), 2007 (Dec.). Pacific Country Report. Sea Level & Climate: Their Present State. Tuvalu, Fig. 15, Monthly sea level at Funafuti, SEAFRAME gauge. *http://www.bom.gov.au/ntc/IDO60033/IDO60033.2007.pdf.*

22. After: Bethke, C.M., 1996. *Geochemical Reaction Modeling.* Oxford University Press, New York, Fig. 6.1 (p. 84).

    Skirrow, G. 1965. The dissolved gases - carbon dioxide. In: Riley, J.H. & Skirrow, G. (Eds.): *Chemical Oceanography.* Academic Press, London, pp. 227-322.

23. After: Liljegren, L., 2008. IPCC Projections overpredict recent warming. The Blackboard. *http://rankexploits.com/musings/2008/ipcc-projections-overpredict-recent-warming/.*

24. After: US Climate Change Science Program (CCSP), 2006. Temperature Trends in the Lower Atmosphere: Steps for Understanding and Reconciling Differences. Final Report, Synthesis and Assessment Product 1.1. *http://www.climatescience.gov/Library/sap/sap1-1/finalreport/.*

25. After: I.E.Frolov et al., 2007. *Scientific research in Arctic. Vol. 2. Climatic changes in the ice cover of the Eurasian shelf seas.* SPb: 'Nauka', p. 113 (Fig. 6.1).

26. After: Akosofu, S.-I., 2009. Two Natural Components of the Recent Climate Change. *http://people.iarc.uaf.edu/~sakasofu/pdf/two_natural_components_recent_climate_change.pdf.*

27. After: Florida State University, 2010. Ryan N. Maue's Seasonal Tropical Cyclone Activity Update. *http://www.coaps.fsu.edu/~maue/tropical/.*

28. After: McIntyre, S., 2009. Sea Ice – June 2009. *http://climate audit.org/2009/07/10/sea-ice-june-2009/.*

29. After: Fisher, D., et al., 2006. Natural Variability of Arctic sea ice over the Holocene . *EOS (Transactions of the American Geophysical Union)* 87 (28), 273, 275, *doi:10.1029/2006EO280001.*

30. After: Intergovernmental Panel on Climate Change (IPCC) 2001. *Climate Change 2001: The Scientific Basis. Contribution of Working Group I to the Third Assessment Report of the IPCC.* Ed.: J.T. Houghton, *et al.,* Cambridge University Press. Fig. 2.18. *http://www.ipcc.ch/ipccreports/tar/wg1/064.htm.*

31. Figure based on information in: Intergovernmental Panel on Climate Change (IPCC) 2001. *Climate Change 2001: The Scientific Basis. Contribution of Working Group I to the Third Assessment Report of the IPCC.* Ed.: J.T. Houghton, *et al.,* Cambridge University Press.

32. After: Cook, E.R., Francey,R.J., Buckley, B.M. & D'Arrigo, R.D., 1996. Long Southern Hemisphere temperature reconstructions from tree rings. *Proceedings of the Royal Society of Tasmania* 130, 65.

33. After: Hansen, J., Fung, I., Lacis, A., Rind, D., Lebedeff, Ruedy, R., Russell, G. & Stone, P. 1988: Global climate changes as forecast by Goddard Institute for Space Studies three-dimensional model. *Journal of Geophysical Research* 93, 9341-9364, *doi:10.1029/88JD00231.*

34. From: Peel, M. C., Finlayson, B. L., & McMahon, T. A., 2007. Updated world map of the Köppen-Geiger climate classification, Hydrol. Earth Syst. Sci. 11, 1633-1644.

Recommended reading and reference material

Many of the references given in the endnotes are to mainstream scientific papers. These papers are important as support for statements made in the main text, but they are not often written with the layperson in mind. Therefore, I provide here a short list of recommended additional books and websites, selected mostly for their readability and accessibility.

1. Introductory background material on climate change
Burroughs, W. (ed.), 2003. *Climate into the 21st Century*. World Meteorological Organization & Cambridge University Press, 240 pp.

A well-illustrated, well-organized and generally well-balanced introduction to the major elements of meteorology and climate change.

Ruddiman, W.F., 2001. *Earth's Climate, Past & Future*. Freeman & Company, New York, 465 pp.

A comprehensive introductory text which covers climate change well and across the board. Contains careful, accurate, well-illustrated and well-balanced explanations of most climate topics.

2. The official (IPCC) view (human-caused climate change is cause for alarm)
Intergovernmental Panel on Climate Change, 2009. *Fourth Assessment Report. Working Group I Report. The Physical Science Basis* (free download or order from http://www.ipcc.ch/ipccreports /ar4-wg1.htm).

Highly technical. The official source of advice on climate change to all governments, worldwide. Summarizes much excellent science, but the more alarmist conclusions are based on many hidden assumptions, and subject to challenge.

3. The independent (NIPCC) review (due-diligence alternative to IPCC's assessment report)

Singer, S.F. & Idso, C., 2009. Climate Change Reconsidered. Nongovernmental International Panel on Climate Change, 880 pp. http://www.nipccreport.org/.

Somewhat technical. But a comprehensive and independent critical assessment of the IPCC's views on climate change. A fundamental reference that includes summaries of many scientific papers that are not taken into account in IPCC reports, and that fails to find evidence for dangerous human-caused global warming.

4. Other books, encompassing the middle-ground view (natural climate change should be planned for; human-caused change, if it emerges, can be adapted to)

Booker, C., 2009. *The Real Global Warming Disaster. Is The Obsession With 'Climate Change' Turning Out To Be The Most Costly Scientific Blunder In History?* Continuum, 368 pp.

The best and most detailed account of the history and socio-political pathology of the global warming issue.

Essex, C. & McKitrick, R., 2007 (2nd ed.). *Taken by Storm. The Troubled Science, Policy and Politics of Global Warming.* Key Porter paperback (available from Amazon CANADA).

Delightfully written, insightful and whimsical account of many of the key issues of global warming science.

Gerhard, L.C. *et al.*, 2001. *Geological Perspectives of Global Climate Change.* American Association of Petroleum Geologists, Studies in Geology #47 (available from AAPG website).

Technical. But all papers in the volume are clearly written, well illustrated and easy to read. Perhaps the best single collection of papers on a wide range of major climate change issues.

Lawson, N., 2007. *An Appeal to Reason: A Cool Look at Global Warming*. Duckworth Overlook, 149 pp.

A short and very well written account, with particular reference to the economic and social aspects of global warming hysteria.

Michaels, P.J. (ed.), 2005. *Shattered Consensus: The True State of Global Warming*. Rowman and Littlefield Publishers, Oxford, 292 pp.

A good introduction to the problems and pitfalls of the global warming debate, with individual chapters written by acknowledged experts in each field.

Nova, J., 2009. *The Skeptics Handbook*. Free download or order from: http://joannenova.com.au/global-warming/.

An attractive, witty and succinct account of the main points in the global warming debate, written in a lively fashion that is suitable for secondary school pupils as well as adults.

Paltridge, G. W., 2009. *The Climate Caper*. Connor Court, 111 pp.

A brief, clearly written and fascinating account of the science and sociology of the global warming phenomenon with especial respect to Australia, written by a former CSIRO Chief Research Scientist.

Plimer, I., 2009. *Heaven And Earth: Global Warming – The Missing Science*. Connor Court, 503 pp.

The geological viewpoint. A profusely referenced account of the history of natural climate change seen through the eyes of a senior research scientist who has long been a doughty warrior for scientific truth.

Spencer, R., 2008. *Climate Confusion. How Global Warming Hysteria Leads to Bad Science, Pandering Politicians and Misguided Policies That Hurt the Poor*. Encounter Books, N.Y., 191 pp.

The meteorological viewpoint. A very readable recounting of important facts about climate change and their implications.

Wishart, I., 2009. Air Con. *The Inconvenient Truth about Global Warming*. Available from: http://briefingroom.typepad.com/the_briefing_room/2009/04/free-preview-of-ian-wisharts-new-book-air-con.html.

Perhaps the most easily readable expose of the climate change debate, and the major lines of science that are involved.

5. Independent climate change websites

This brief list does not include the major public database and reference sites on climate such as NASA, NOAA, CRU,national meteorological offices, etc., which are widely known and easily found through web search engines. Listed here instead are some selected less obvious, but independent and critical, sources of information.

D'Aleo, J. – IceCap – *http://icecap.us/index.php*.
Humlum, O. – Climate4you – *http://www.climate4you.com/*.
Idso, C. & S. – CO_2 Science – h*ttp://www.co2science.org/*.
International Climate Science Coalition –
http://www.internationalcsc.org.au/.
McIntyre, S. – Climate Audit – *http://climateaudit.org/*.
Nova, J. – JoNova – *http://joannenova.com.au/*.
Science and Public Policy Institute (SPPI) –
http://scienceandpublicpolicy.org/.
Watts, A. – Watts Up With That – *http://wattsupwiththat.com/*.

Chapter notes

1   The term 'global warming', as popularly used, is shorthand for the cumbersome phrase 'dangerous global warming caused by human carbon dioxide emissions'. For brevity and readability, I will sometimes adopt that usage in this book. Similarly, many people use 'climate change' as a synonym for 'global warming', with the same implicit definition of dangerous human causation. I will prefer to use 'climate change' in its native, self-evident meaning, adding the qualifiers 'human-caused' or 'natural' where necessary.

2   Stern, N., 2006. *Stern Review on the Economics of Climate Change*, H.M. Treasury, *http://www.hmtreasury.gov.uk/independent_reviews/stern_review _economics_climate_change/sternreview_index.cfm*.

3   Nordhaus, W., 2007. *The Challenge of Global Warming: Economic Models and Environmental Policy*, Yale University, New Haven, Connecticut, USA. *http://nordhaus.econ.yale.edu/dice_mss_072407_all.pdf*.

4   Garnaut, R., 2008. *The Garnaut Climate Change Review*. Cambridge University Press, 680 pp. *http://www.garnautreview.org.au/domino/Web_ Notes/Garnaut/garnautweb.nsf*.

5   Intergovernmental Panel on Climate Change (IPCC) 1990. *Climate Change: The IPCC Scientific Assessment. Report Prepared for IPCC by Working Group I.* Ed.: J.T. Houghton, *et al.*, Cambridge University Press.

     Intergovernmental Panel on Climate Change (IPCC) 1992. *Climate Change 1992: The Supplementary Report to the IPCC Scientific Assessment.* Ed.: J.T. Houghton, *et al.*, Cambridge University Press.

     Intergovernmental Panel on Climate Change (IPCC) 1996. *Climate Change 1995: The Science of Climate Change.* Ed.: J.T. Houghton, *et al.*, Cambridge University Press.

     Intergovernmental Panel on Climate Change (IPCC) 2001. *Climate Change 2001: The Scientific Basis. Contribution of Working Group I to the Third Assessment Report of the IPCC.* Ed.: J.T. Houghton, *et al.*, Cambridge University Press.

     Intergovernmental Panel on Climate Change (IPCC) 2007. *Climate Change 2007: The Physical Science Basis. Contribution of Working Group*

I to the Fourth Assessment Report of the IPCC. Ed,: Susan Solomon, *et al.*, Cambridge University Press.

6   Nova, J., 2008. The Skeptics Handbook II. Global Bullies Want Your Money. *http://joannenova.com.au/2009/11/skeptics-handbook-ii-global-bullies-want-your-money/*. This paper identifies only the money spent from the US public purse on global warming matters. Meanwhile, other governments have also been committing significant sums, plus various industry lobby groups, and especially the large environmental NGOs. In total, more than \$100 billion must have been spent pursuing the global warming scare since 1990.

7   Carter, R.M., De Freitas, C.R., Goklany, I.M., Holland, D. & Lindzen, R.S., 2007. Climate change. Climate science and the Stern Review. *World Economics* 8, 161-82.

      Holland, D., Carter, R.M., de Freitas, C.R., Goklany, I.M. & Lindzen, R.S., 2007 Climate change. Response to Simmonds and Steffen. *World Economics* 8, 143-51.

      Tol, R.S.J., 2006. The Stern Review of the economics of climate change: a comment. In: 'The Stern Report: Some Early Criticisms', Center for Science & Public Policy, p. 32-8. *http://www.ff.org/centers/csspp/pdf/20061104_stern.pdf*.

8   McLean, J., 2007, August/September. An Analysis of the Review of the IPCC 4AR WG I Report. Science & Public Policy Institute. *http://mclean.ch/climate/docs/IPCC_review_updated_analysis.pdf*.

      McLean, J., 2007, November. Why the IPCC Should be Disbanded. *http://scienceandpublicpolicy.org/originals/whytheipccshouldbedisbanded.html*.

      McLean, J., 2009. The IPCC Can't Count its 'Expert Scientists' – Author and Reviewer Numbers are Wrong, International Climate and Environmental Change Assessment Project. *http://mclean.ch/climate/docs/IPCC_numbers.pdf*.

      McLean, J., 2008. Prejudiced Authors, Prejudiced Findings. Analysis of IPCC data on chapter authors and reviewers published through the Science and Public Policy Institute. Particularly pages 16-17. *http://www.heartland.org/policybot/results/23573/Prejudiced_Authors_Prejudiced_Findings.html*.

9   Essex, C. & McKitrick, R., 2002. *Taken by Storm. The Troubled Science, Policy and Politics of Global Warming*. Key Porter Books, Toronto, p. 12.

10   Essex, C. & McKitrick, R., ibid, p. 305.

11   Perlwitz, J., Hoerling, M., Eischeid, J., Xu, T. & Kumar, A., 2009. A strong bout of natural cooling in 2008. *Geophysical Research Letters* 36, L23706, doi:10.1029/2009GL041188.

12 Easterling, D. R. & Wehner, M.F., 2009. Is the climate warming or cooling? *Geophysical Research Letters* 36, L08706, *doi:10.1029/2009GL037810*.

13 Douglass, D.H., Christy, J.R., Pearson, B.J. & Singer, S.F., 2007. A comparison of tropical temperature trends with model predictions. *International Journal of Climatology, doi:10.1002/doc.1651*.

14 Though this statement remains true as I write, in mid-January, 2010, a very marked swing towards global warming scepticism was underway in public and media opinion across the Western world.

15 DEFRA, 2005. Symposium on Avoiding Dangerous Climate Change. Exeter, 1-3 Feb. 2005. *http://www.stabilisation2005.com/*.

16 The United Nations Framework Agreement on climate Change. *http://unfccc.int/2860.php*.

17 NASA, 2010 (25 Jan.). On the Shoulders of Giants. Milutin Milankovitch (1879-1958). NASA Earth Observatory. *http://earthobservatory.nasa.gov/Features/Milankovitch/*.

   Hays, J.D., Imbrie, J. & Shackleton, N.J., 1976. Variations in the Earth's Orbit: Pacemaker of the Ice Ages. *Science* 194, 1121-32.

18 Berger, A. & Loutre, M.F., 1991. Insolation values for the climate of the last 10 million years. *Quaternary Science Reviews* 10, 297-317.

19 Davis, B.A.S., Brewer, S., Stevenson, A.C. & Guiot, J., 2003. The temperature of Europe during the Holocene reconstructed from pollen data. *Quaternary Science Reviews* 22, 1701-16, *doi:10.1016/S0277-3791(03)00173*.

   Williams, P.W., King, D.N.T., Zhao, J.-X. & Collerson, K.D., 2004. Speleothem master chronologies: combined Holocene ^{18}O and ^{13}C records from the North Island of New Zealand and their paleoenvironmental interpretation. *The Holocene* 14, 194-208, *doi:10.1191/0959683604hl676rp*.

   Kaufman, D.S., *et al.*, 2004. Holocene thermal maximum in the western Arctic (0-18° W). *Quaternary Science Reviews* 23, 529-60, *doi:10.1016/j.quascirev.2003.09.007*.

20 Emiliani, C., 1955. Pleistocene temperatures. *Journal of Geology* 63, 538-78.

   Shackleton, N.J. & Opdyke, N.D., 1973. Oxygen isotope and palaeo-magnetic stratigraphy of equatorial Pacific core V28-238 : oxygen isotope temperatures and ice volumes in a 10^5 and 10^6 year scale. *Quaternary Research* 3, 39-55.

21 Dansgaard,W., Johnsen, S.J.,Moller, J., Langway Jr., C.C., 1969. One

thousand centuries of climate record from Camp Century on the Greenland Ice Sheet. *Science* 166, 377-81.

NGRIP – The North Greenland Ice Core Project, 2009. *http://www.glaciology.gfy.ku.dk/ngrip/index_eng.htm.*

Watanabe, O., Jouzel, J., Johnsen, S., Parrenin, F., Shoji, H. & Yoshida, N., 2003. Homogeneous climate variability across East Antarctic over the past three glacial cycles. *Nature* 422, 509-12.

Augustin, L., Barbante, C., Barnes, P.R., *et al.*, 2004. Eight glacial cycles from an Antarctic ice core. *Nature* 429, 623-8. *doi:10.1038/nature02599. PMID 15190344.*

22   Mudelsee, M., 2001. The phase relations among atmospheric CO_2 content, temperature and global ice volume over the past 420 ka. *Quaternary Science Reviews* 20, 583-9.

23   Taylor, K.C., Lamorey, G.W., Doyle, G.A., Alley, R.B., Grootes, P.M., Mayewski, P.A., White, J.W.C. & Barlow, L.K., 1993. The flickering switch of late Pleistocene climate change. *Nature* 361, 432-5.

Steffensen, J.P. *et al.*, 2008. High-resolution Greenland ice core data show abrupt climate change happens in a few years. *Science Express*, 19 June 2008/10.1126/science.1157707.

Brauer *et al.*, 2008. An abrupt wind shift in western Europe at the onset of the Younger Dryas cold period. *Nature Geoscience* 1(8): 520 *doi: 10.1038/ngeo263.*

Lie, O. & Paasche, O., 2006. How extreme was northern hemisphere seasonality during the Younger Dryas? *Quaternary Science Reviews* 24, 1159-82.

24   Broecker, W.S., 2006. Abrupt climate change revisited. *Global and Planetary Change. doi:10.1016/j.gloplacha.2006.06.01*

25   Bond, G., Kromer, B., Beer, J., Muscheler, R., Evans, M.N., Showers, W., Hoffmann, S., Lotti-Bond, R., Hajdas, I. & Bonani, G., 2001. Persistent solar influence on North Atlantic climate during the Holocene. *Science* 294, 2130-6.

Skilbeck, C.G., Rolph, T.C., Hill, N., Woods, J. & Wilkens, R.H., 2005. Holocene millenial/centennial-scale multiproxy cyclicity in temperate eastern Australian estuary sediments. *Journal of Quaternary Science* 20, 327-47.

Moros, M. *et al.*, 2009. Holocene climate variability in the Southern Ocean recorded in a deep-sea sediment core off South Australia. *Quaternary Science Reviews* 28, 1932-40.

Willard, D.A., Bernhardt, C.E., Korejwo, D.A. & Meyers, S.R., 2005.

Impact of millenial-scale Holocene climate variability on eastern North American terrestrial ecosystems: pollen-based climate reconstruction. *Global & Planetary Change* 47, 17-35.

Avery, D.T. & Singer, S.F., 2008 (2nd Ed.). *Unstoppable Global Warming: Every 1,500 Years*. Rowman & Littlefield Publishers.

26   Soon, W. & Baliunas, S., 2003. Proxy climatic and environmental changes of the past 1000 years. *Climate Research*, 23, 89-110.

Idso, C., 2010. Mediaeval Warm Period Project. *http://www.co2science.org/data/mwp/mwpp.php*.

27   Loehle, C., 2007. A 2000-year global temperature reconstruction based on non-tree ring proxies. *Energy and Environment* 18, 1049-58.

Loehle, C. & McCulloch, J.H., 2008. Correction to: A 2000-year global temperature reconstruction based on non-tree ring proxies. *Energy and Environment* 19, 93-100.

28   Sclater, J.G., Jaupart, C. & Galson, D., 1980. The Heat Flow Through Oceanic and Continental Crust and the Heat Loss of the Earth. *Reviews of Geophysics & Space Physics* 18, 269-311.

29   Meehl, G.A., Arblaster, J.M., Matthes, K., Sassi, F. & van Loon, H., 2009. Amplifying the Pacific Climate System Response to a Small 11-Year Solar Cycle Forcing. *Science* 325, 1114-8, *doi: 10.1126/science.1172872*.

Soon, W.H., Posmentier, E.S. & Baliunas, S.L., 1996. Inference of solar irradiance variability from terrestrial temperature changes, 1880-1993: An astrophysical application of the sun-climate connection. *The Astrophysical Journal* 472, 891-902.

Soon, W.H., 2005. Variable solar irradiance as a plausible agent for multi-decadal variations in the Arcticwide surface air temperature record of the past 130 years. *Geophysical Research Letters* 32, *doi:10.1029/2005GL023429*.

30   Alexander, W.J.R., 2005. Linkages between Solar Activity and Climatic Responses. *Energy & Environment* 16, 239-53 *http://www.ingentaconnect.com/content/mscp/ene/2005/00000016/00000 002/art00003*.

31   Lassen, K., 2009. Long-term variations in solar activity and their apparent effect on the Earth's climate, 15 pp. *http://www.tmgnow.com/repository/solar/lassen1.html*.

Cini Castagnoli, G., Taricco, C. & Alessio, S., 2005. Isotopic record in a marine shallow-water core: imprint of solar centennial cycles in the past 2 millennia. *Advances in Space Research* 35, 504-8.

32   Lambeck, K., & Cazenave, A., 1976. Long-term variations in the
      length of day and climatic change. *Geophysical Journal of the Royal
      Astronomical Society* 46, 555-73.

33   Yasuda, I., 2009. The 18.6-year period moon-tidal cycle in Pacific
      Decadal Oscillation reconstructed from tree-rings in western North
      America. *Geophysical Research Letters* 36, L05605,
      doi:10.1029/2008GL036880.

      Yndestad H., Turrell W.R. & Ozhigin, V., 2008. Lunar nodal tide
      effects on variability of sea-level, temperature, and salinity in the
      Faroe-Shetland Channel and the Barents Sea. *Deep Sea Research I* 55,
      1201-17.

34   Agnihotri, R. & Dutta, K., 2003. Centennial scale variations in
      rainfall (Indian, east equatorial and Chinese monsoons):
      Manifestations of solar variability. *Current Science* 85, 459-63.

      Liu, J., Wang, B., *et al.*, 2009. Centennial variations of the global
      monsoon precipitation in the last millennium: Results from ECHO-G
      model. *Journal of Climate* 22, 2356-71.

35   The concept of forcing in the climate system is intimately tied to the
      presumption of radiative balance at the top of the atmosphere, i.e. that
      the incoming energy from the sun is equivalent to the outgoing heat
      from the Earth plus the energy absorbed by planetary processes.
      Though clearly not the case over geological time-scales, radiative
      balance is nonetheless a useful starting point from which to examine
      modern climate. Agents that upset the presumed radiative balance,
      and therefore change the climate, are termed climate *forcings*. Forcings
      may be external to the Earth, such as changes in the radiative intensity
      of the sun, or internal, such as changes in cloud cover (which changes
      the average reflectivity, or albedo, of the planet), aerosol dust (which
      may be derived from either volcanic or human sources) or greenhouse
      gases (such as human carbon dioxide emissions).

      An initial change produced by a climate forcing, such as an increase in
      temperature caused by an increase in solar radiation, is usually
      moderated by various *feedback* effects, which may be positive or
      negative. For example, an increase in carbon dioxide will cause an
      initially small increase in temperature that then itself causes more
      evaporation from the ocean. Water vapour being a greenhouse gas, its
      increase results in the positive feedback effect of a further temperature
      increase; at the same time, the evaporation creates more low clouds
      which (because they reflect solar radiation back into space) comprise a
      negative feedback loop that causes cooling. The balance of all the
      feedbacks that are involved in the climate system is unknown, and
      perhaps unknowable. The lack of knowledge regarding feedback effects

is one of the major reasons that climate GCMs are likely to remain unreliable far into the future.

36 Svensmark, H., Olaf, J., Pedersen, P., Marsh, N., Enghoff, M. & Uggerhøj, U., 2007. Experimental Evidence for the role of ions in particle nucleation under atmospheric conditions. *Proceedings of the Royal Society of London* A463, 385-96.

37 Neff, U. *et al.*, 2001. Strong coherence between solar variability and the monsoon in Oman between 9 and 6 kyr ago. *Nature* 411, 290-3.

38 Dansgaard, W., Johnsen, S.J., Clausen, H.B. & Langway Jr., C.C., 1971. Climate record revealed by the Camp Century ice core. In: Turekian, K.K. (Ed.), *Late Cenozoic Glacial Ages Symposium*. Yale UP, pp. 37-56.

    Rahmstorf, S., 2003. Timing of abrupt climate change: A precise clock. *Geophysical Research Letters*, 30, 10, 1510, *doi:10.1029/2003GL017115*.

39 Keeling, C.D. & Whorf, T.P., 2000. The 1,800-year oceanic tidal cycle: a possible cause of rapid climate change. *Proceedings of the US National Academy of Sciences* 97, 3814-19.

40 Marshall Space Flight Center, NASA, 2010. Solar Cycle Prediction (Updated 2010/01/04). *http://solarscience.msfc.nasa.gov/predict.shtml*.

41 Friis-Christensen, E. & Lassen, K., 1991. Length of the Solar Cycle: An Indicator of Solar Activity Closely Associated with Climate. *Science* 254, 698–700, *doi: 10.1126/science.254.5032.698*.

    Archibald, D., 2006. Solar Cycles 24 & 25 and predicted climate response. *Energy and Environment* 17, 29-38.

    Archibald, D., 2007. Climate outlook to 2030. *Energy and Environment* 18, 615-9.

    Butler, C.J. & Johnston, D.J., 1996. A provisional long mean air temperature series for Armagh Observatory. *Irish Astronomical Journal* 21, 251-73.

    Clilverd, M.A., Clarke, E., Ulrich, T., Rishbeth, H. & Jarvis, M.J., 2006. Predicting Solar Cycle 24 and beyond. Space *Weather 4*, S09005, *doi:10.1029/2005SW000207. http://www.pdfdownload.org/pdf2 html/pdf2html.php?url=http%3A%2F%2Fusers.telenet.be%2Fj.janssens% 2FSC24Clilverd.pdf&images=yes*.

42 Archibald, D., 2010 (3 Feb.). Solar Cycle 24 Update. *http://wattsupwiththat.com/2010/02/02/solar-cycle-24-update/*.

43 Watts, A., 2010 (7 Jan.). Solar geomagnetic index reaches unprecedented low – only 'zero' could be lower – in a month when sunspots became more active.

http://wattsupwiththat.com/2010/01/07/suns-magneticindex-reaches-unprecedent-low-only-zero-could-be-lower-in-a-month-when-sunspots-became-more-active/.

44 Archibald, D., 2009. NASA now saying that a Dalton Minimum repeat is possible. 29 July 2009, Solar Cycle 25. *http://solarcycle25.com/index.php?id=51*.

45 Soon, W., 2009. Solar Arctic-mediated climate variation on multi-decadal to centennial timescales: Empirical evidence, mechanistic explanation, and testable consequences. *Physical Geography* 30, 144-84.

46 Hingane, L.S., 1996. Is a signature of socio-economic impact written on the climate? *Climatic Change* 32, 91-102.

McKitrick, R. & McMichaels, P. 2004. A test of corrections for extraneous signals in gridded surface temperature data. *Climate Research* 26, 159-73.

McKitrick, R.R. & Michaels, P.J., 2007. Quantifying the influence of anthropogenic surface processes and inhomogeneities on gridded global climate data. *Journal of Geophysical Research* 112, D24S09, *doi:10.1029/2007JD008465*, 2007.

Brandsma, T., Konnen, G.P. & Wessels, H.R.A., 2003. UHI effects. *International Journal of Climatology* 23, 829-45.

47 Steyaert, L.T. & Knox, R.G., 2008. Reconstructed historical land cover and biophysical parameters for studies of land-atmosphere interactions within the eastern United States. *Journal of Geophysical Research* 113, D02101, *doi:10.1029/2006JD008277*.

Fall, S., Niyogi, D., Gluhovsky, A, Pielke Sr., R.A., Kalnay, E. & Rochon, G., 2009. Impacts of land use and land cover on temperature trends over the continental United States: Assessment using the North American Regional Reanalysis. *International Journal of Climatology, doi: 10.1002/joc.1996*.

Pielke Sr., R.A. *et al.*, 2007. Unresolved issues with the assessment of multi-decadal global land surface temperature trends. *Journal of Geophysical Research* 112, D24S08, *doi:10:1029/2006JD008229*.

48 Carter, R.M., De Freitas, C.R., Goklany, I.M., Holland, D. & Lindzen, R.S., 2007. Climate change. Climate science and the Stern Review. *World Economics* 8, 161-82.

Pielke, R.A., 2002. Overlooked issues in the US National Climate and IPCC Assessment: an editorial essay. *Climatic Change 52*, 1-11.

Cooling: The Human Climate Signal? A Note from 'Cohenite'. *http://www.jennifermarohasy.com/blog/archives/003303.html*.

49 Brignell, J., 2009. A complete list of the things caused by global warming. *Number Watch*, 11 Dec. 2009. *http://www.numberwatch.co.uk/warmlist.htm*.

50 Hulme, M., Dessai, S., Lorenzoni, I. & Nelson, D.R., 2009. Unstable climates: Exploring the statistical and social constructions of 'normal' climate. *Geoforum* 40, 197-206. *doi:10.1016/j.geoforum.2008.09.010*.

51 Manley, G., 1974. Central England Temperatures: monthly means 1659 to 1973. *Quarterly Journal of the Royal Meteorological Society* 100, 389-405.

52 Met Office (UK), 2010. HadCET: Central England Temperature. *http://www.metoffice.gov.uk/climatechange/science/monitoring/hadcet.html*; *http://badc.nerc.ac.uk/data/cet/*.

53 Flood, W., 2009a. Britain's temperature and rainfall. *http://mclean.ch/climate/England_Scotland.htm*.

    Flood, W., 2009b. No global warming in 351 year British temperature record. The Carbon Sense Coalition. *http://carbon-sense.com/2009/10/01/british-record/*.

54 MET Office (UK), 2010. Climate monitoring and datasets. *http://www.metoffice.gov.uk/climatechange/science/monitoring/*.

55 NASA (Goddard Institute for Space Studies – GISS), 2010. GISS Surface Temperature Analysis. *http://data.giss.nasa.gov/gistemp/*.

56 Akasofu, S., 2009. Natural causes of 20th Century warming: recovery from the Little Ice Age and oscillatory change. Heartland-2 International Conference on Climate Change, New York, 9 March 2009. *http://www.heartland.org/events/NewYork09/proceedings.html*.

    Akasofu, Syun-Ichi, 2009. Two natural components of the recent climate change: (i) The recovery from the Little Ice Age; (ii) The Multi-decadal Oscillation. *http://people.iarc.uaf.edu/~sakasofu/pdf/two_natural_components_recent_climate_change.pdf*.

57 McKitrick, R.R. & Michaels, P.J., 2008. Quantifying the influence of anthropogenic surface processes and inhomogeneities on gridded global climate data.

    McKendry, I.G., 2003. Progress Report: Applied Climatology. *Progress in Physical Geography* 27, 597-606.

58 Brohan, P., Kennedy, J., Tett, S., Harris, I. & Jones, P., 2005. Report on HadCRUT3 including error estimates. MS—RAND—CPP—PROG0407. *http://stratasphere.com/blog/wpcontent/uploads/hadcrut3_gmr_defra_report_200503.pdf*.

59   Balling, R.C. Jr., Idso, S.B. & Hughes, W.S., 1992. Long-Term and Recent Anomalous Temperature Changes in Australia. *Geophysical Research Letters* 19, 2317-20.

See also other unpublished 1990s papers that establish that a strong urban heat island effect is present in Australian temperature data, at: Hughes, W., 2009 (10 Dec.). Some rare balance on the taxpayer funded ABC for a change. *http://www.warwickhughes.com/blog/?tag=urban-heat-island.*

60   D'Aleo, J. & Watts, A., 2010. Surface Temperature Records: Policy Driven Deception? SPPI Original Paper, 26 Jan. 2010. *http://scienceandpublicpolicy.org/images/stories/papers/originals/surface_tem p.pdf.*

61   Thorne, P.W., Parker, D.E., Tett, S.F.B., Jones, P.D., McCarthy, M., Coleman, H. & Brohan, P., 2005. Revisiting radiosonde upper air temperatures from 1958 to 2002, *Journal of Geophysical Research 110, D18105, doi:10.1029/2004JD005753.*

62   RSS - *http://wattsupwiththat.wordpress.com/2008/02/04/rss-satellite-data-for-jan08-2nd-coldest-januaryfor-the-planet-in-15-years/;*

UAH - *http://wattsupwiththat.wordpress.com/2008/02/06/uah-satellite-data-forjan08-in-agreement-with-rss-data/.*

63   Gray, V, 2006. Temperature trends in the lower atmosphere. *Energy & Environment* 17, 707-14.

64   Loehle, C., 2009. Trend analysis of RSS and UAH MSU global temperature data. *Energy & Environment* 20, 1087-98.

65   Briggs, W.M., 2010. Actually, Weather *Is* Climate. Pajamas Media, 22 Jan. 2010. *http://pajamasmedia.com/blog/actually-weather-is-climate/.*

66   Hurrell, J., Meehl, G.A., Bader, D., Delworth, T.L., Kirtman, B., & Wielicki, B., 2009. A unified modeling approach to climate system prediction. *Bulletin of the American Meteorological Society* 90, 1819-32, *doi: 10.1175/2009BAMS2752.1.*

67   Davis, J.C. & Bohling, G.C., 2001. The search for patterns in ice-core temperature curves. In: Gerhard, L.C. *et al.* (eds.), *Geological Perspectives of Global Climate Change, American Association of Petroleum Geologists, Studies in Geology,* 47, 213-29.

68   Wong, P., 2010. Climate action can't be put off. *The Australian*, 2 Feb. 2010. *http://www.theaustralian.com.au/news/opinion/climate-action-cant-be-put-off/story-e6frg6zo-1225825672648.*

69   Ring, Ed, 2008. Has global warming alarm become the goal rather than the result of scientific research? Is climate science really designed to answer questions? *EcoWorld, Climate Science.*

http://www.ecoworld.com/global-warming/climate-science-is-it-currently-designed-to-answer-questions.html.

70   Sundquist, E.T., 1985. Geological perspectives on carbon dioxide and the carbon cycle. *In* Sundquist, E.T. & Broecker, W.S. (Eds.): The carbon cycle and atmospheric CO_2: natural variations Archean to present. *American Geophysical Union, Geophysical Monograph* 32, 5-59.

Segalstad, T. V., 1998. Carbon cycle modelling and the residence time of natural and anthropogenic atmospheric CO_2: on the construction of the 'Greenhouse Effect Global Warming' dogma. In: Bate, R. (Ed.): *Global warming: the continuing debate.* ESEF, Cambridge, UK [ISBN 0952773422]: 184-219. Available at: *http://www.co2web.info/ESEF3VO2.pdf.*

Segalstad, T.V., 2009. Correct timing is everything – also for CO_2 in the air. *CO_2 Science* 12, no. 31, 5 Aug. 2009. *http://www.co2web.info/Segalstad_CO2-Science_090805.pdf.*

Segalstad, T.V., 2009. Carbon isotope mass balance modeling of atmospheric vs. oceanic CO_2. 2nd Heartland *International Conference on Climate Change*, New York, 8-10 March, Track 1 Paper 1. *http://www.heartland.org/bin/media/newyork09/PowerPoint/Tom_Segalstad .ppt* - see slide 14.

71   Barrett, J., 2005. Greenhouse molecules, their spectra and function in the atmosphere. *Energy & Environment* 16, 1037-45. *http://www.warwickhughes.com/papers/barrett_ee05.pdf.*

72   Baliunas, S. & Soon, W., 1999. Pioneers in the Greenhouse Effect. *World Climate Report* 4, *http://www.worldclimatereport.com/archive/ previous_issues/vol4/v4n19/cutting.htm.*

73   Kiehl, J.T. & Trenberth, K.I., 1997. Earth's Annual Global Mean Energy Budget. *Bulletin of the American Meteorological Society* 78, 197-208.

74   Segalstad, T. V., 1996. The distribution of CO_2 between atmosphere, hydrosphere, and lithosphere; minimal influence from anthropogenic CO_2 on the global 'Greenhouse Effect'. *In* Emsley, J. (Ed.): The Global Warming Debate. The Report of the European Science and Environment Forum. Bourne Press Ltd., Bournemouth, Dorset, UK (ISBN 0952773406), pp. 41-50. *http://www.co2web.info/ESEFVO1.pdf*

75   de Freitas, C.R., 2002. Are observed changes in the concentration of carbon dioxide in the atmosphere really dangerous? *Bulletin of Canadian Petroleum Geology* 50, 297-327.

Soon, W., 2007. Implications of the secondary role of carbon dioxide and methane forcing in climate change: past present and future. *Physical Geography* 28, 97-125.

76   Jaworowski, Z., Segalstad, T.V. & Hisdal, V., 1992. Atmospheric CO_2 and global warming: a critical review; 2nd revised edition. *Norsk Polarinstitutt, Meddelelser [Norwegian Polar Institute, Memoirs]* 119, 76 pp. *http://www.co2web.info/np-m-119.pdf.*

Jaworowski, Z., Segalstad, T.V. & Ono, N., 1992. Do glaciers tell a true atmospheric CO_2 story? *Science of the Total Environment* 114, 227-84. *http://www.co2web.info/stoten92.pdf.*

77   Slocum, G., 1955. Has the amount of carbon dioxide in the atmosphere changed significantly since the beginning of the twentieth century? *Monthly Weather Review* October (1955), 225-31.

78   Beck, E.-G., 2007. 180 years of atmospheric CO_2 gas analysis by chemical methods. *Energy & Environment* 18, 259-82.

79   Gerlach, T.M. *et al.*, 2001. CO_2 degassing at Kilauea Volcano: implications for primary magma, summit reservoir dynamics, and magma supply monitoring. *American Geophysical Union, Fall Meeting, Abstract #V22E-10.*

80   Lupton, J., *et al.*, 2006. Submarine venting of liquid carbon dioxide on a Mariana Arc volcano. *Geochemistry, Geophysics, Geosystems* 7, Q08007, *doi:10.1029/2005GC001152. http://dx.doi.org/10.1029/2005GC001152.*

'Although CO_2 is generally the most abundant dissolved gas found in submarine hydrothermal fluids, it is rarely found in the form of CO_2 liquid. Here we report the discovery of an unusual CO_2-rich hydrothermal system at 1600-m depth near the summit of NW Eifuku, a small submarine volcano in the northern Mariana Arc. The site, named Champagne, was found to be discharging two distinct fluids from the same vent field: a 103°C gas-rich hydrothermal fluid and cold (<4°C) droplets composed mainly of liquid CO_2. The hot vent fluid contained up to 2.7 moles/kg CO_2, the highest ever reported for submarine hydrothermal fluids. The liquid droplets were composed of 98% CO_2, 1% $H2S$, with only trace amounts of $CH4$ and $H2$. Surveys of the overlying water column plumes indicated that the vent fluid and buoyant CO_2 droplets ascended <200 m before dispersing into the ocean. The discovery of such a high CO_2 flux at the Champagne site, estimated to be about 0.1% of the global MOR carbon flux, suggests that submarine arc volcanoes may play a larger role in oceanic carbon cycling than previously realized.'

81   Plimer, I., 2009. *Heaven and Earth: Global Warming – The Missing Science.* Connor Court, 503 pp.

82   Robinson, A.B., Robinson, N.E. & Soon, W., 2007. Environmental effects of increased atmospheric carbon dioxide. *Journal of American Physician and Surgeons* 12, 79-90. *http://www.jpands.org/vol12no3/robinson600.pdf.*

83 Oelkers, E.H. & Cole, D.R., 2008. Carbon dioxide sequestration: a solution to a global problem. *Elements* 4, 305-10.

84 Ball, T., 2008. Environmentalists seize green moral high ground ignoring science. *Canada Free Press*, 13 June 2008. *<http://canadafreepress.com/index.php/article/3482>*.

Common estimates of carbon dioxide emissions are:

Respiration (humans, animals, phytoplankton)	43.5-52 Gt C/ year
Ocean Outgassing (tropical areas)	90-100 Gt C/year
Soil Bacteria, decomposition	50-60 Gt C/ year
Volcanoes, soil degassing	0.5-2 Gt C/ year
Forest cutting, forest fires	0.6-2.6 Gt C/year
Anthropogenic emissions (2005)	7.2-7.5 Gt C/year
TOTAL	192-224 Gt C/ year
ERROR	32 Gt/C (~15%)

The range of estimates of natural and human carbon dioxide production in 2005 is 192-224 Gt C/year (Gigatonnes of carbon per year), with an uncertainty of 32 Gt. The human contribution of 7.5 Gt lies within the error range of the first three natural sources, and the total error range is almost 5 times the human production.

85 Lindzen, R., 2006. Understanding common climate claims, in *Proceedings International Seminar on Nuclear War and Planetary Emergencies* (World Federation of Scientists). *http://www.climatescience.org.nz/assets/20060507_O_Lindzen.pdf*.

86 Lindzen, R.S., 2009. The Climate Science isn't settled: confident predictions of catastrophe are unwarranted. *Wall Street Journal, Opinion,* 30 Nov. *http://online.wsj.com/article/SB10001424052748 703939404574567423917025400.html*.

87 Chylek, P., Lohmann, U., Dubey, M., Mishchenko, M., Kahn, R. & Ohmura, A., 2007. Limits on climate sensitivity derived from recent satellite and surface observations. *Journal of Geophysical Research 112,* D24S04, *doi:10.1029/2007JD008740*.

Douglass, D.H., Christy, J.R., Pearson, B.J. & Singer, S.F., 2007. A comparison of tropical temperature trends with model predictions. *International Journal of Climatology, doi: 10.1002/doc.1651*.

Douglass, D.H. & Knox, R.S., 2005. Climate forcing by the volcanic eruption of Mount Pinatubo. *Geophysical Research Letters 32,* L05710, *doi:10.1029/2004GL022119*.

Idso, S.B., 1998. CO_2-induced global warming: a skeptic's view of potential climate change. *Climate Research* 10, 69-82.

Kärner, O., 2002. On nonstationarity and antipersistency in global temperature series. *Journal of Geophysical Research* 107, *doi:10.1029/2001JD002024*.

Kärner, O., 2005. Some examples of negative feedback in the Earth climate system. *Central European Journal of Physics* 3, 190-208.

Lindzen, R.S., Chou, M.-D. & Hou, A.Y., 2001. Does the Earth have an adaptive infrared iris? *Bulletin of the American Meteorological Society* 82, 417-32.

Lindzen, R.S., Chou, M.-D. & Hou, A.Y., 2002. Comments on 'No evidence for iris.' *Bulletin of the American Meteorological Society* 83, 1345-8.

Lindzen, R.S. & Choi, Y-S., 2009. On the determination of climate feedbacks from ERBE data. *Geophysical Research Letters 36, L16705, doi:10.1029/2009GL039628*.

Monckton, C., 2008. Climate Sensitivity Reconsidered. American Physical Society, Forum on Physics & Society. *http://www.aps.org/units/fps/newsletters/200807/monckton.cfm*.

Paltridge, G., Arking, A. & Pook, M., 2009. Trends in middle- and upper-level tropospheric humidity from NCEP reanalysis data. *Theoretical and Applied Climatology DOI 10.1007/s00704-009-0117-x*.

Schwartz, S.E., 2007. Heat capacity, time constant, and sensitivity of Earth's climate system. *Journal of Geophysical Research 112, D24S05, doi:10.1029/2007JD008746*.

Schwartz, S.E., 2008. Reply to comments by G. Foster *et al.*, R. Knutti *et al.*, and N. Scafetta on 'Heat capacity, time constant, and sensitivity of Earth's climate system'. Schwartz S. E. *Journal of Geophysical Research, 113, D15105 (2008), doi:10.1029/2008JD009872*.

Spencer, R.W. & Braswell, W.D., 2008. Potential biases in feedback diagnosis from observations data: a simple model demonstration. *Journal of Climate* 21, 5624-8.

Schwartz, S., 2007. Heat capacity, time constant and sensitivity of Earth's climate system. *Journal of Geophysical Research*, DOI 10.1029/2007JD008746.

Soon, W., 2005. Variable solar irradiance as a plausible agent for multi-decadal variations in the Arctic-wide surface air temperature record of the past 130 years. *Geophysical Research Letters*, vol. 32, doi:10.1029/2005GL023429.

Soon, W., 2009. Solar Arctic-mediated climate variation on multi-decadal to centennial timescales: Empirical evidence, mechanistic explanation, and testable consequences. *Physical Geography*, vol. 30, 144-84.

89   Idso, C., Idso, S.B. & Balling, R.C., 2001. An intensive two-week
      study of an urban CO_2 dome in Phoenix, Arizona, USA. *Atmospheric
      Environment* 35, 995-1000. *doi:10.1016/S1352-2310(00)00412-X*.

90   CO_2 Science, 2009. Urban CO_2 dome (Phoenix, Arizona, USA) –
      Summary. *http://www.co2science.org/subject/u/summaries/
      phxurbanco2dome.php*.

91   Francey, R.J., Allison, C.E., Etheridge, D.M., Trudinger, C.M., Enting,
      I.G., Leuenberger, M., Langenfelds, R.L., Michel, E. & Steele, L.P.,
      1999. A 1000-year high precision record of ^{13}C in atmospheric CO_2.
      Tellus 51B, 170-93.

92   Keeling, Charles D., *et al.*, 1989. A Three-Dimensional Model of
      Atmospheric CO_2 Transport Based on Observed Winds. In *Aspects of
      Climate Variability in the Pacific and the Western Americas (AGU
      Monograph 55)*, edited by David H. Peterson, pp. 165-363. American
      Geophysical Union, Washington, DC.

93   Keeling, R.F., Piper, S.C. & Heimann, M., 1996. Global and
      hemispheric sinks deduced from changes in atmosphere O2
      concentration. *Nature* 381, 218-21.

94   Bard, E., 2005. The source of historic increases in atmospheric carbon
      dioxide. Letter, *Physics Today* (May 2005), 16-7.

95   Archer, D., *et al.*, 2007. Atmospheric lifetime of fossil fuel carbon
      dioxide. *Annual Review of Earth and Planetary Sciences* 37, 117-34,
      doi:10.1146/annurev.earth.031208.100206.

96   Lam, H.S.H., 2003. Residence Time of Atmospheric CO_2. 4 pp.,
      http://www.princeton.edu/~lam/TauL1b.pdf.

97   Segalstad, T.V., 1992. The amount of non-fossil-fuel CO_2 in the
      atmosphere. *American Geophysical Union, Chapman Conference on
      Climate, Volcanism, and Global Change, 23-27 March, 1992, Hilo,
      Hawaii. Abstracts*, 25.

      Segalstad, T.V., 1996. CO_2 and climate – are we fumbling with the
      Earth's thermostat? *Norwegian Oil Review* 22 (10), p. 297.

98   Spencer, R.W., 2009 (21 Jan.). Increasing Atmospheric CO_2:
      Manmade…or Natural? *http://www.drroyspencer.com/2009/01/
      increasing-atmospheric-co2-manmade%E2%80%A6or-natural/*.

      Spencer, R.W., 2009 (15 May). Global Warming Causing Carbon
      Dioxide Increases: A Simple Model. *http://www.drroyspencer.
      com/2009/05/global-warming-causing-carbon-dioxide-increases-a-
      simple-model/*.

99   European Project for Ice Coring in Antarctica (EPICA), 2010.
      http://www.esf.org/index.php?id=855.

North Greenland Ice Core Project (NGRIP), 2010. *http://www.gfy.ku.dk/~www-glac/ngrip/index_eng.htm.*

100 Stott, L., Timmermann, A. & Thunell, R., 2007. Southern Hemisphere and Deep-Sea Warming Led Deglacial Atmospheric CO_2 Rise and Tropical Warming. *Science Express*, 27 Sept. 2007, 1-4, *doi: 10.1126/science.1143791.*

101 Kuo, C., Lindberg, C. & Thomson, D.J., 1990. Coherence established between atmospheric carbon dioxide and global temperature. *Nature* 343, 709-13.

102 Kouwenberg, L., WagnAqer, R., Kurschner, W. & Visscher, H., 2005. Atmospheric CO_2 fluctuations during the last millennium reconstructed by stomatal frequency analysis of *Tsuga heterophylla* needles. *Geology* 33, 33-6.

Wagner, F., Kouwenberg, L.L.R., van Hoof, T.B. & Visscher, H., 2004. Reproducibility of Holocene atmospheric CO_2 records based on stomatal frequency. *Quaternary Science Reviews* 23, 1947-54.

103 Haworth, M., 2005. Mid-Cretaceous pCO_2 based on stomata of the extinct conifer *Pseudofrenelopsis* (Cheirolepidiaceae). *Geology* 33, 749-52.

Veizer, J., Godderis, Y. & François, L.M., 2000. Evidence for decoupling of atmospheric CO_2 and global climate during the Phanerozoic eon. *Nature* 408, 698-701, *doi: 10.1038/35047044.*

104 Eamus, D., 1996. Responses of field grown trees to CO_2 enrichment, *Commonwealth Forestry Review* 75, 39-47.

Saxe, H., Ellsworth, D. S. & Heath, J., 1998. Tree and forest functioning in an enriched CO_2 atmosphere. *New Phytologist* 139, 395-436.

Wittwer, S., 2005. *Food, Climate & Carbon Dioxide: The Global Environment and World Food Production*, Boca Raton, Florida: CRC Press.

105 McMahon, S.M., Parker, G.G., & Miller, D.R., 2010. Evidence for a recent increase in forest growth. *Proceedings of the National Academy of Sciences. info:/10.1073/pnas.0912376107.*

106 Tubiello, F.N., *et al.*, 2007. Crop response to elevated CO_2 and world food supply. A comment on 'Food for Thought...' by Long *et al.*, *Science* 312:1918–1921, 2006. *European Journal of Agronomy* 26, 215-23.

Asseng, S. *et al.*, 2009. *Crop Physiology*, Chapter 20, pp. 511-43.

107 Xiao, J. and Moody, A., 2005. Geographical distribution of global greening trends and their climatic correlates: 1982-1998. *International Journal of Remote Sensing* 26, 2371-90.

108 West, L., 2007. US Supreme Court Rejects Bush Policy on Vehicle Greenhouse Gas Emissions. *About.com: Environmental Issues*. *http:// environment.about.com/od/environmentallawpolicy/a/epa_greenhouse.htm*.

109 Krauthammer, C., 2009. The new socialism, *The Patriot Post* (Chattanooga, Tennessee), 11 December 2009. *http://patriotpost.us/opinion/charles-krauthammer/2009/12/11/the-new-socialism/*.

110 NOAA, Office of Climate Observation (OCO), 2010. Role of the Ocean in Climate (accessed Jan 23., 2010). *http://www.oco.noaa.gov /index.jsp?show_page=page_roc.jsp&nav=universal*.

111 Castles I., and Henderson, D., 2003. The IPCC Emission Scenarios: An Economic-Statistical Critique. *Energy & Environment 14, 159-85*.

Castles, I., 2008. Economic formulas in IPCC report criticized for overstating emissions. *Environment & Climate News*, The Heartland Institute, 1 March 2008. *http://www.globalwarmingheartland.com/ Article.cfm?artId=22786*.

McKibbin, W.J., Pearce, D. & Stegman, A., 2009. Climate change scenarios and long-term projections. *Climatic Change, doi: 10.1007/s10584-009-9621-3*.

112 Singer, S.F. & Idso, C., 2009. Climate Change Reconsidered. Nongovernmental International Panel on Climate Change, 880 pp. *http://www.nipccreport.org/*.

113 *The Australian Baseline Sea-level Monitoring Project, Annual Sea-level Data Summary Report*. National Tidal Centre, Bureau of Meteorology. *http://www.bom.gov.au/oceanography/projects/abslmp/reports_yearly. shtml*.

114 Holgate, S.J., 2007. On the decadal rates of sea-level change during the twentieth century. *Geophysical Research Letters* 34, L01602, *doi:10.1029/2006GL028492*.

115 Wöppelmann, G., *et al.*, 2010. Rates of sea-level change over the past century in a geocentric reference frame. *Geophysical Research Letters* 36, L12607, *doi:10.1029/2009GL038720*.

116 University of Colorado at Boulder, 2010. Sea-level change (accessed 24 Jan. 2010). *http://sealevel.colorado.edu/*.

117 Wunsch, C., Ponte, R.M. & Heimback, P., 2007. Decadal trends in sea-level patterns: 1993-2004. *Journal of Climate* 20, 5889-5911. *doi: 10.1175/2007JCLI1840.1*.

118 Since 2003, ocean temperatures have been cooling along with the atmosphere (Willis *et al.*, 2007; 2009; Loehle, 2009). Indeed, Harrison & Carson (2007), studying ocean temperature change between 1950

and 2000, have even suggested that ocean cooling may have been occurring since the late 1970s, though the databases around which these and other similar papers are constructed are far from ideal. Another recent study by Carson & Harrison (2008) showed that previously inferred warmings of the oceans were largely an artefact of the data interpolation schemes used, which were biased towards the 30 per cent of the ocean that warmed and ignored regions that cooled. This immediately casts doubt, for instance, on the claims by IPCC (2007) that extra heat has accumulated in the oceans since 1960.

Carson, M. & Harrison, D.E., 2008. Is the Upper Ocean Warming? Comparisons of 50-Year Trends from Different Analyses. *Journal of Climate* 21(10), 2259-68.

Harrison, D.E. & Carson, M., 2007. Is the World Ocean Warming? Upper-Ocean Temperature Trends: 1950-2000. *Journal of Physical Oceanography* 37, 174-87.

Loehle, C., 2009. Cooling of the global ocean since 2003. *Energy & Environment* 20, 101-4.

Willis, J.K., *et al.*, 2007. Correction to 'Recent cooling of the upper ocean.' *Geophysical Research Letters* 34, 16, doi:10.1029/2007GL030323.

Willis, J.K., Lyman, J.M., Johnson, G.C. & Gilson, J., 2009. In-situ data biases and recent ocean heat content variability. *Journal of Atmospheric and Oceanic Technology* 26, 846-52.

119   Hannah, J., 2004. An updated analysis of long-term sea-level change in New Zealand. *Geophysical Research Letters* 31, doi: 10.1020/2003GL019166.

120   Summarized in Appendix 1 of Carter, R.M., 2008. Some notes on sea-level change around the Australian coastline. Article 136 at *http://members.iinet.net.au/~glrmc/new_page_1.htm*.

121   Church, J. A. & White, N. J., 2006. A 20th century acceleration in global sea-level rise. *Geophysical Research Letters* 33, L01602, doi:10.1029/2005GL024826.

Burton, D.A., 2010. Analysis of global linear mean sea-level (MSL) trends, including distance-weighted averaging. http://www.burtonsys.com/climate/global_msl_trend_analysis.html.

122   Church, J.A., White, N.J., Coleman, R., Lambeck, K. & Mitrovica, J.X., 2004. Estimates of the regional distribution of sea-level rise over the 1950-2000 period. *Journal of Climate* 17, 2609-25.

123   Center for Research on Environmental Decisions, Columbia University, 2009. The Psychology of Climate Change Communication. *http://www.cred.columbia.edu/guide/*

124 Inman, M., 2009. Where Warming Hits Hard. *Nature Reports Climate Change*. Published online 15 Jan. 2009, *doi:10.1038/climate.2009.3*. *http://www.nature.com/climate/2009/0902/full/climate.2009.3.html#B6*.

125 Bureau of Meteorology, 2009. South Pacific Sea-level and Climate Monitoring Program. *http://www.bom.gov.au/pacificsealevel/index.shtml* and *http://www.bom.gov.au/ntc/IDO60102/IDO60102.2009_1.pdf*.

126 Priestley, C.H.B., 1966. The limitation of temperature by evaporation in hot climates. *Agricultural Meteorology* 3, 241-6.

127 Eschenbach, W., 2004. Tuvalu Not Experiencing Increased Sea-level Rise. *Energy & Environment* 15, 527-543, *doi: 10.1260/0958305041494701*.

128 Eschenbach, W., 2010 (27 Jan.). Floating islands. *http://wattsupwiththat.com/2010/01/27/floatingislands/#more-15762*.

129 ABC News, 5 Aug. 2008. Climate change ruling threatens coastal property prices. *http://www.abc.net.au/news/stories/2008/08/05/2324468.htm*.

   Evans, I., 2009. Byron Bay to be abandoned to the waves. Crikey, 8 July 2009. *http://www.crikey.com.au/2009/07/08/byron-bay-to-be-abandoned-to-the-waves*.

130 Kump, L.R., Brantley, S.L. & Arthur, M.A., 2000. Chemical weathering, atmospheric CO_2 and climate. *Annual Review of Earth and Planetary Sciences* 28, 611-67.

131 Walker, J.B., Hays, P.B. & Kasting, J.F., 1981. A negative feedback mechanism for the long-term stabilization of the Earth's surface temperature. *Journal of Geophysical Research* 86, 9776-82.

   A buffered solution is one which resists changes in pH when small quantities of an acid or an alkali are added to it. This happens because materials in the solution react with, and remove, the added hydrogen or hydroxide ions that would otherwise result in a change in pH.

132 Pearson, P.N. & Palmer, M.R., 1999. Middle Eocene seawater pH and atmospheric carbon dioxide concentrations. *Science* 284, 1824-6, *doi: 10.1126/science.284.5421.1824*.

133 Holland, H.D., 1984. *The chemical evolution of the atmosphere and oceans*. Princeton University Press, 582 pp.

134 Lemarchand, D., Gaillardet, J., Lewin, É. & Allègre, C.J., 2000. The influence of rivers on marine boron isotopes and implications for reconstructing past ocean pH. *Nature* 408, 951-4.

135 Skirrow, G., 1965. The dissolved gases - carbon dioxide. In Riley, J.P. & Skirrow, G. (Eds.). Chemical Oceanography. Academic Press, London, pp. 227-322.

136   Revelle, R. & Fairbridge, R., 1957. Carbonates and carbon dioxide. *Geological Society of America, Memoir* 67, 239-85.

137   Pelejero, C., Eva Calvo, C., McCulloch, M.T., Marshall, J.F., Gagan, M.K., Lough, J.M. & Opdyke, B.N., 2005. Preindustrial to modern interdecadal variability in coral reef pH. *Science* 309, 2204-7, *doi: 10.1126/science.1113692.*

      See also: Comment by Matear, R.J. & McNeil, B.I., 2006. *Science 314,* 595b, *doi: 10.1126/science.1128198;* and Reply by Pelejero *et al.*, 2006. *Science* 314, 595c, *doi: 10.1126/science.1128502.*

138   Liu, X., *et al.*, 2009. Instability of seawater pH in the South China Sea during the mid-late Holocene: evidence from boron isotopic composition of corals. *Geochimica et Cosmochimica Acta* 73, 1264-72, *doi:10.1016/j.gca.2008.11.034 .*

139   Henry's Law was formulated in 1803 by the English chemist William Henry, and applies to dilute solutions and low gas pressures. The law states that at a constant temperature, the amount of a given gas dissolved in a liquid is directly proportional to the partial pressure of the gas in equilibrium with the liquid.

140   Orr, *et al.*, 2005. Anthropogenic ocean acidification over the twenty-first century and its impact on calcifying organisms. *Nature* 437, 681–686, *doi:10.1038/nature04095.*

141   Guinotte, J.M., Buddemeier, R.W. & Kleypas, J.A., 2003. Future coral reef habitat marginality: temporal and spatial effects of climate change in the Pacific basin. *Coral Reefs* 22, 551-8.

142   Raven, J., *et al.*, 2005. *Ocean Acidification Due to Increasing Atmospheric Carbon Dioxide.* Policy Document 12/05, The Royal Society of London, 57 pp. *http://www.royalsoc.ac.uk/document.asp?id=3249.*

      Guinotte, J.M. & Fabry, V.J., 2008. Ocean acidification and Its potential effects on marine ecosystems. *Annals New York Academy of Sciences* 1134, 320-42. *doi: 10.1196/annals.1439.01*

      Wright, S. & Davidson, A., 2006. Ocean acidification: a newly recognized threat. *Australian Antarctic Magazine* 10, 26-27. *http://www.aad.gov.au/MediaLibrary/asset/MediaItems/ml_38895483703 7037_17%20Ocean%20acidification.pdf.*

143   Bellamy, D. & Barrett, J., 2005. No Proof Yet. Letter, Oct. 2005. *Chemistry World*, p. 27.

144   Idso, C., 2010. 'Acid Test: the Global Challenge of Ocean Acidification' – a new propaganda film by the National Resources Defense Council fails the acid test of real world data. A critique. *Science & Public Policy Institute Original Paper*, 5 Jan. 2010, 54 pp.

Segalstad, T.V., 2008. Carbon isotope mass balance modelling of atmospheric vs. oceanic CO_2. 33rd International Geological Congress, Norway, 8 Aug. 2008, Program & Abstracts. *https://abstracts.congrex.com /scripts/JMEvent/ProgrammeLogic_Abstract_P.asp?PL=Y&Form_Id=8&Client_Id='CXST'&Project_Id='08080845'&Person_Id=1345952.*

Segalstad, T.V., 2009. Carbon Isotope Mass Balance Modeling of Atmospheric vs. Oceanic CO_2. 2nd Heartland International Conference on Climate Change, New York, 10 March 2009, Session I, Track 1: Palaeoclimatology. *http://www.heartland.org/events/ NewYork09/proceedings.html.*

145   Nielsdottir, M.C., Moore, M.C., Sanders, R., Hinz, D.J. & Achterberg, E.P., 2009. Iron limitation of the postbloom phytoplankton communities in the Iceland Basin. *Global Biogeochemical Cycles* 23, GB3001, *doi:10.1029/2008GB003410.*

146   NASA Goddard Space Flight Centre, 2010. SeaWIFS Project. *http://oceancolor.gsfc.nasa.gov/SeaWiFS/.*

147   Casey, N. & McClain, C., 2005. Satellites See Ocean Plants Increase, Coasts Greening. NASA, Goddard Space Flight Center, 2 March 2005. *http://www.nasa.gov/centers/goddard/news/topstory/chlorophyll.html.*

148   Moy, A.D., Howard, W.R., Bray, S.G. & Trull, T.W. , 2009. Reduced calcification in modern Southern Ocean planktonic foraminfera. *Nature Geoscience* 2, 276-80, *doi: 10.1038/ngeo460.*

Roberts, D., *et al.*, 2008. Interannual variability of pteropod shell weights in the high-CO_2 Southern Ocean. *Biogeosciences Discussions* 5, 4453-80.

149   Iglesias-Rodriguez, M.D., *et al.*, Phytoplankton calcification in a high-CO_2 world. *Science* 320, 336-40. See also: *http://sciencewatch.com/dr/fbp/2009/09aprfbp/09aprfbpRod/.*

150   Szmant, A., 2009. In: Pennisi, E., Calcification rates drop in Australian reefs. *Science* 323, 27. *doi: 10.1126/science.323.5910.27.*

151   Krauskopf, K.B., 1979. *Introduction to Geochemistry* (2nd ed._ McGraw-Hill, 617 pp.

152   Hendriks, L.E., Duarte, C.M. & Alvarez, M. 2010. Vulnerability of marine biodiversity to ocean acidification: a meta-analysis. *Estuarine, Coastal and Shelf Science* 86, 157-64.

Pichler, T. & Dix, G.R., 1996. Hydrothermal venting within a coral reef ecosystem, Ambitle Island, Papua New Guinea. *Geology* 24, 435-8. *doi: 10.1130/0091-7613(1996).*

Pichler, T., Heikoop, J.M., Risk, M.J., Veizer, J. & Campbell, I.L., 2000. Hydrothermal effects on isotope and trace element records in modern reef corals: a study of *Porites lobata* from Tutum Bay, Ambitle Island, Papua New Guinea. *Palaios* 15, 225-34. *doi: 10.1669/0883-1351(2000)015*.

153   Herfort, L., Thake, B. & Taubner, I., 2008. Bicarbonate stimulation of calcification and photosynthesis in two hermatypic corals. *Journal of Phycology* 44, 91-8.

154   Byrne, M., *et al.*, 2009. Temperature, but not pH, compromises sea urchin fertilization and early development under near-future climate change scenarios. *Proceedings of the Royal Society* B276, 1883-1888, *doi: 10.1098/rspb.2008.1935*.

155   Tunnicliffe, V., *et al.*, 2009. Survival of mussels in extremely acidic waters on a submarine volcano. *Nature Geoscience*, 12 April 2009, *doi: 10.1038/NGEO500*.

156   Lough J.M. & Barnes D.J., 2000. Environmental controls on growth of the massive coral *Porites*. *Journal of Experimental Marine Biology and Ecology* 245, 225-43.

157   McIntyre, S., 2009. 'Unprecedented' in the past 153 Years. *Climate Audit*, 3 June 2009. *http://climateaudit.org/2009/06/03/unprecedented-in-at-least-the-past-400-years/*.

Ridd, P.V. *et al.*, 2010. A critique of a method to determine long-term decline of coral reef ecosystems. *In preparation*.

158   Atkinson, M.J., Carlson, B.A. & Crow, G.L., 1995. Coral growth in high-nutrient, low-pH seawater: a case study of corals cultured at the Waikiki Aquarium, Honolulu, Hawaii. *Coral Reefs* 14, 215-23.

159   Landesman, P., 2009. The Mathematics of Global Warming. American Thinker, 8 Dec. 2009. *http://www.americanthinker.com/2009/11/the_mathematics_of_global_warm.html*.

160   Zillman, J., 2003. World Meteorological Address. John Zillman, Director of the Australian Bureau of Meteorology, 21 March 2003. http://www.bom.gov.au/ents/media_releases/ho/20030320a.shtml.

'The most important question – should global warming proceed as the IPCC reports suggest – is how will warming be manifest at the national, regional and local level, and what would that mean for each of us? I believe this question is, at present, completely unanswerable.'

MacCracken, M., Smith, J. & Janetos, A.C., 2004. Reliable regional climate model not yet on horizon. *Nature* 429, 699.

'We strongly agree that much more reliable regional climate simulations and analyses are needed. However, at present … such simulations are more aspiration than reality.'

161 Reid, J., 2009. Climate Modeling Nonsense. *Quadrant Online*, October 2009. *http://www.quadrant.org.au/magazine/issue/2009/10/climate-modelling-nonsense*.

162 Stainforth, D.A., *et al.*, 2005. Uncertainty in predictions of the climate response to rising levels of greenhouse gases. *Nature* 433, 403-6.

163 Zhang, M.H., *et al.*, 2005. Comparing clouds and their seasonal variations in 10 atmospheric general circulation models with satellite measurements. *Journal of Geophysical Research* 110: D15SO2, *doi:10.1029/2004JD005021*.

164 This is known as the Kelvin Fallacy, after the famous British physicist who in 1862 in a paper entitled 'on the Secular Cooling of the Earth' miscalculated the age of the Earth as 100 million years, using the first principles of physics. The error occurred because at the time scientists had no understanding that the Earth had a second internal heat source as well as its molten core, namely the radioactive decay of elements. The fallacy was the assumption that the physics of the system was fully understood.

165 Soon, W., Baliunas, S., Idso, S.B., Kondratyev, K.Y. & Posmentier, E.S. 2001. Modeling Climatic Effects of Anthropogenic Carbon Dioxide Emissions: Unknowns and Uncertainties. *Climate Research* 18, 259-75.

166 Essex, C., 2009. Fundamental Uncertainties in Climate Modeling. In: McKitrick, R. (ed.), *Supplementary Analysis of the Independent Summary for Policymakers*, Fraser Institute, 99-114. *www.fraserinstitute.org /commerce.web/product_files/CriticalTopicsInGlobalWarming.pdf*.

167 Liljegren, Lucia, 2008. IPCC Projections Overpredict Recent Warming. The Blackboard. *http://rankexploits.com/musings/2008/ipcc-projections-overpredict-recent-warming/*.

168 Keenlyside, N.S., Latif, M., Jungclaus, J., Kornblueh, L. & Roeckner, E., 2008. Advancing decadal-scale climate prediction in the North Atlantic sector. *Nature* 453, 84-8, *doi:10.1038/nature06921*.

169 Paltridge, G., Arking, A. & Pook, M., 2009. Trends in middle- and upper-level tropospheric humidity from NCEP reanalysis data. *Theoretical and Applied Climatology, doi: 10.1007/s00704-009-0117-x*.

170 Wentz, F.J., Ricciardulli, L., Hilburn, K. & Mears, C., 2007. How much more rain will global warming bring? *Science* 317, 233-5.

171 Koutsoyiannis, D., Efstratiadis, D. A., Mamassis, N. & Christofides, A., 2008. On the credibility of climate predictions. *Hydrological Sciences–Journal–des Sciences Hydrologiques* 53, 471-84.

172 Thompson, D.W.J., Kennedy, J.J., Wallace J.M. & Jones, P.D., 2008. A large discontinuity in the mid-twentieth century in observed global-mean surface temperature. *Nature*, 453, 29 May 2008,

doi:10.1038/nature06982.

173 *http://www-argo.ucsd.edu/rey_line_atlas.gif; http://www-argo.ucsd.edu/nino3_4_atlas.gif; http://www.argo.ucsd.edu/Marine_Atlas.html.*

174 Loehle, C., 2009. Cooling of the global ocean since 2003. *Energy & Environment* 20 (1&2), 101-4.

    Tisdale, R., 2010. OHC Linear Trends and Recent Update of NODC OHC (0-700 Meters) Data. Climate Observations, 5 Feb. 2010. *http://bobtisdale.blogspot.com/2010/02/ohc-linear-trends-and-recent-updateof.html.*

175 DiPuccio, W., 2009. The global warming hypothesis and ocean heat. *http://wattsupwiththat.com/2009/05/06/the-global-warming-hypothesis-and-ocean-heat/.*

176 Halide, H. & Ridd, P., 2008. Complicated ENSO models do not significantly outperform very simple ENSO models. *International Journal of Climatology* 28, 219-33, *doi: 10.1002/joc.1519.*

177 Lawrence, P., 2009. Scientists Tackle Climate Model Mystery. *http://www.nsf.gov/discoveries/disc_summ.jsp?cntn_id=114910&org=NSF.*

    Lawrence, P. J. & Chase, T.N., 2010. Investigating the Climate Impacts of Global Land Cover Change in the Community Climate System Model (CCSM 3.0). *International Journal of Climatology, 10.1002/joc.2061.*

178 Trenberth, K.E., 2007. Predictions of climate. *Nature - Climate Feedback Blog,* 4 June 2007. *http://blogs.nature.com/climatefeedback/2007/06/predictions_of_climate.html.*

179 Kotov, S.R., 2001. Near-term climate prediction using ice-core data from Greenland, in Geological Perspectives of Global Climate Change (eds: L.C. Gerhard *et al*), American Association of Petroleum Geologists, *Studies in Geology* 47:305-15.

180 Klyashtorin, L.B. & Lyubushin, A.A., 2003. On the coherence between dynamics of the world fuel consumption and global temperature anomaly. *Energy & Environment* 14, 733-82.

181 Dr Willem de Lange (*pers. comm.*) explains the PDO in the following terms: '*The ENSO cycle involves heat stored in the upper 500-700 m of the ocean. The PDO involves heat stored at 500-1500 m in the ocean, i.e. at depths closer to the main thermocline. Due to the greater depth and slower circulation, the PDO has a longer "periodicity" than ENSO, though neither oscillation has a true fixed period.*

    '*The PDO in operation can be viewed as a climate storage heater, with El Nino's representing extra heat discharge and La Nina's representing extra heat storage. These changes are superimposed on the normal seasonal inflow/outflow of heat, and comprise a lagged oscillator within the ocean/atmosphere system that responds to the inflow of solar radiation, which itself also varies because of changing solar activity and amounts of cloudiness.'*

182 Loehle, C., 2004. Climate change: detection and attribution of trends from long-term geologic data. *Ecological Modelling* 171: 433-50.

183 Lin Zhen-Shan & Sun Xian, 2007. Multi-scale analysis of global temperature changes and trend of a drop in temperature in the next 20 years. *Meteorology & Atmospheric Physics* 95, 115-21, *doi: 10.1007/s00703-006-0199-2.*

184 *http://www.forecastingprinciples.com/, http://www.forecasters.org/, http://www.forecasters.org/ijf/index.php*

185 Green, K.C., & Armstrong, J.S., 2007. Global warming: Forecasts by scientists versus scientific forecasts. *Energy & Environment* 18, 997-1022.

     Green, K., Armstrong, S. & Soon, W., 2009. Validity of climate change forecasting for public policy decision making. *International Journal of Forecasting* 25, 826-32. *http://kestencgreen.com/gas-2009-validity.pdf.*

186 Roberts, D., 2006. An interview with accidental movie star Al Gore. GRIST, 9 May 2006. *http://www.grist.org/article/roberts2/.*

187 Johnston, I., 2007. Man who wants to turn every Scots child into an environmental evangelist. *The Scotsman,* 17 Jan. 2007. *http://findarticles.com/p/news-articles/scotsman-edinburgh-scotlandthe/mi_7951/is_2007_Jan_17/turn-scots-child-environment-evangelist/ai_n34578972/.*

188 BBC – One minute world news, 11 Oct. 2007. Gore climate film's nine 'errors'. *http://news.bbc.co.uk/2/hi/uk_news/education/7037671.stm.*

189 England and Wales High Court (Administrative Court) Decisions. 10 Oct. 2007. Judgement between Stewart Dimmock and Secretary of State for Education and Skills. *http://www.bailii.org/ew/cases/EWHC/Admin/2007/2288.html.*

190 Monckton, C., 2007. 35 Inconvenient Truths. The Errors in Al Gore's Movie. SPPI Paper, 21 pp. *http://scienceandpublicpolicy.org/images/stories/press_releases/monckton-response-to-gore-errors.pdf.*

     See also: Lewis, Marlo, Jr., 2007. Al Gore's Science Fiction: A Skeptic's Guide to *An Inconvenient Truth.* Competitive Enterprise Institute, Congressional Working Paper. *http://cei.org/pdf/5820.pdf.*

191 Mudelsee, M., 2001. The phase relations among atmospheric CO_2

content, temperature and global ice volume over the past 420 ka. *Quaternary Science Reviews* 20, 583-9.

Vakulenko, N.V., Kotlyakov, V.M., Monin, A.S. & Sonechkin, D.M., 2004. Evidence for the leading role of temperature variations relative to greenhouse gas concentration variations in the Vostok ice core record. *Doklady Earth Sciences* 397, 663-7.

192 Francey, R. J., 1999. A 1000-year high precision record of $\delta^{13}C$ in atmospheric CO_2. *Tellus B* 51, 170-93.

Quay, P. D., Tilbrook, B. & Wong, C. S., 1992. Oceanic uptake of fossil fuel CO_2: carbon-13 evidence. *Science* 256, 74-9.

193 Ashok, K., Guan, Z. & Yamagata, T., 2003. Influence of the Indian Ocean Dipole on the Australian winter rainfall. *Geophysical Research Letters* 30, 1821, *doi:10.1029/2003GL017926*.

Cai, W., Cowan, T. & Raupach, M., 2009. Positive Indian Ocean Dipole events precondition southeast Australia bushfires. *Geophysical Research Letters* 36, L19710, *doi:10.1029/2009GL039902*.

194 Wong, P., 2009. Letter to Senator Fielding, 18 June 2009, as commented on in item 7 at: *http://joannenova.com.au/global-warming/the-wong-fielding-meeting-on-global-warming-documents/*.

The Bureau of Meteorology has stated that '*the combination of record heat and widespread drought during the past five to ten years over large parts of southern and eastern Australia is without historical precedent and is, at least partly, a result of climate change*' (BOM Drought Statement, 3 July 2008).

195 Lockart, N., Kavetski, D. & Franks, S.W., 2009. On the recent warming in the Murray Darling Basin – land surface interactions misunderstood. *Geophysical Research Letters* 36, L24405, *doi:10.1029/2009GL040598*. See also *http://www.newcastle.edu.au/news/2009/11/flawsrevealedinclimatechangeresearch.html*.

196 Klotzbach, P.J., 2006. Trends in global tropical cyclone activity over the past twenty years (1986-2005). *Geophysical Research Letters* 33, L10805, doi: 1029/2006GL025881.

Maue, R.N., 2009a. Global tropical cyclone accumulated cyclone energy, 24 month running sums, through July 2009. *http://www.coaps.fsu.edu/~maue/tropical/*.

Maue, R.N., 2009b. Northern hemisphere tropical cyclone activity. *Geophysical Research Letters* 36, L05805, *doi:10.1029/2008GL035946*.

197 Nott, J., Haig, J., Neil, H. & Gillieson, D., 2007. Greater frequency of variability of landfalling tropical cyclones at centennial compared to

seasonal and decadal scales. *Earth & Planetary Science Letters* 255, 367-72.

Nott, J., Smithers, S., Walsh, K. & Rhodes, E., 2009. Sand beach ridges record 6000 year history of extreme tropical cyclone activity in northeastern Australia. *Quaternary Science Reviews* 28, 1511-20.

198 World Meteorological Organization (WMO), 2006. Consensus statement on hurricanes and global warming. WMO Commission on Atmospheric Sciences, Tropical Meteorology Research Program Panel. *http://www.bom.gov.au/info/CAS-statement.pdf*.

199 Muir-Wood, R., Miller, S. & Boissonade, A., 2006. *The Search for Trends in a Global Catalogue of Normalised Weather-Related Catastrophe Losses*, Climate Change and Disaster Losses Workshop, Hohenkammer.

Pielke, Jr., R. A., Gratz, J., Landsea, C. W., Collins, D., Saunders, M. & Musulin, R., 2008. Normalized Hurricane Damages in the United States: 1900-2005. *Natural Hazards Review* 9, 29-42.

Zhang, Q., Wu, L. & Liu, Q., 2009. Tropical cyclone damages in China 1983-2006. *Bulletin of the American Meteorological Society* 90, 489-95.

200 Polyakov, I.V., *et al.*, 2009. Variability and trends of air temperature and pressure in the maritime Arctic, 1875 – 2000. *Journal of Climate* 16 (12), 2067-77.

201 Klyashtorin, L.B. & Lyubushin, A.A., 2003. On the Coherence between Dynamics of the World Fuel Consumption and Global Temperature Anomaly. *Energy & Environment* 14, 773-82.

Frolov, I.E., Gudkovich, Z.M., Karklin, V.P., Kovalev, Ye.G. & Smolyanitsky, V.M., 2007. *Scientific research in Arctic. Vol. 2. Climatic changes in the ice cover of the Eurasian shelf seas.* Nauka, St. Petersburg, pp. 1-158.

202 Chylek, P., Box, J.E. & Lesins, G., 2004. Global warming and the Greenland ice sheet. *Climatic Change* 63, 201-21.

Przybylak, R., 2000. Temporal and spatial variation of surface air temperature over the period of instrumental observations in the Arctic. *International Journal of Climatology* 20, 587-614.

203 Comiso, J.C., 2000. Variability and trends in Antarctic surface temperatures from in *situ* and satellite infra red measurements. *Journal of Climate* 13, 1674-96.

Doran, P.T., *et al.*, 2002. Antarctic climate cooling and terrestrial ecosystem response. *Nature*, 13 Jan. 2002, *doi:10/1038/nature710.*

204   Davis, C.H., Li, Y., McConnell, J.R., Frey, M.M. & Hanna, E., 2005. Snow-driven growth in East Antarctica icesheet mitigates recent sea-level rise. *Science* 308, 1898-1901.

Johannessen, O.M., Khvorostovsky, K., Miles, M.W. & Bobylev, L.P., 2005. Recent ice-sheet growth in the interior of Greenland. *Science Express*, 20 October 2005.

Wingham, D.J., Shepperd, A., Muir, A. & Marshall, G.J., 2006. Mass balance of the Antarctic ice sheet. Philisophical Transactions of the Royal Society A364, 1627-35.

205   Chen, J., *et al.*, 2009. Accelerated Antarctic ice loss from satellite gravity measurements. *Nature Geoscience* 2, 859-62.

206   Krinner, G., *et al.*, 2007. Simulated Antarctic precipitation and surface mass balance at the end of the twentieth and twenty-first centuries. *Climate Dynamics* 28, 215-30.

Zwally, H.J., *et al.*, 2005. Mass changes of the Greenland and Antarctic ice sheets and shelves and contributions to sea-level rise: 1992-2002. *Journal of Glaciology* 51, 509-27.

207   Chapman, W. L.,2010. The Cryosphere Today. *http://arctic.atmos.uiuc.edu/cryosphere* - accessed 24/01/2010.

208   Fisher, D., Dyke, A., Koerner, R., Bourgeois, J., Kinnard, C., Zdanowicz, C., de Vernal, A., Hillaire-Marcel, C., Savelle, J. & Rochon, A., 2006. Natural variability of Arctic sea ice over the Holocene. *EOS, Transactions AGU* 87 (No. 29), 273.

209   Ollier, C. & Pain, 2009. Why the Greenland and Antarctic Ice Sheets are Not Collapsing. *Australian Institute of Geologists, News* (August 2009). *http://wattsupwiththat.com/2009/08/27/why-the-greenland-andantarctic-ice-sheets-are-not-collapsing/*.

210   12 more glaciers that haven't heard the news about global warming. 25 Jan. 2010. *http://www.ihatethemedia.com/12-more-glaciers-that-havent-heard-the-news-about-global-warming*.

211   US Geological Survey, 2010. Sea-level and Climate. *http://pubs.usgs.gov/fs/fs2-00/*.

212   NOAA, National Snow & Ice Center, 2010. World Glacier Inventory. *http://nsidc.org/data/g01130.html*.

213   World Glacier Monitoring Service, University of Zurich, 2010. Preliminary glacier mass balance data 2007/2008. *http://www.geo.unizh.ch/wgms/mbb/sum08.html*.

214   Zemp, M., Hoelzle, M. & Haeberli, W. 2009. Six decades of glacier mass-balance observations: a review of the worldwide monitoring

network. *Annals of Glaciology* 50, 101-11.

215 Joerin, U.E., Stocker, T.F. & Schlüchter, C., 2006. Multicentury glacier fluctuations in the Swiss Alps during the Holocene. *The Holocene* 16, 697-704, *doi: 10.1191/0959683606hl964rp*.

216 Nicolussi, K. & Patzelt, G., 2000. Discovery of early-Holocene wood and peat on the forefield of the Pasterze Glacier, Eastern Alps, Austria. *The Holocene* 10, 191-9.

217 Mote, P. W. & Kaser G., 2007. The shrinking glaciers of Kilimanjaro: can global warming be blamed? *American Scientist* 95, 318-25.

Duane, W.J., Pepin, N.C., Losleben, M.L. & Hardy, D.R., 2008. General characteristics of temperature and humidity variability on Kilimanjaro, Tanzania. *Arctic, Antarctic, and Alpine Research* 40, 323-34.

218 IUCN Species Survival Commission, 2001. Polar Bears: Proc. 13th Working Meeting, IUCN/SCC Polar Bear Specialist Group 23-28 June 2001, Nuuk, Greenland. *http://pbsg.npolar.no/docs/PBSG13proc.pdf*.

219 Aars, J., Lunn, N. & Derocher, A. (eds.), 2006. Polar bears. *Proceedings of the 14th Working Meeting of the IUCN/SCC Polar Bear Specialist Group*, 20-24 June 2005, Seattle, Wash.. IUCN, Gland, Switzerland & Cambridge.

220 McLean, J., 2010 (9 Feb.). Sea-surface Temperatures along the Great Barrier Reef. *http://mclean.ch/climate/GBR_sea_temperature.htm*.

221 Carter, R.M., 2006. Great news for the Great Barrier Reef: Tully River water quality. *Energy & Environment* 17(4), 527-48.

222 Starck, W., 2005. 'Threats' to the Great Barrier Reef. Institute of Public Affairs, Backgrounder, 21 pp. *http://www.ipa.org.au/publications /614/%27threats%27-to-the-great-barrier-reef*.

223 Crichton, M., 2003. Environmentalism as religion. Commonwealth Club San Francisco, 15 Sept. 2003. *http://www.crichton-official.com/speech-environmentalismaseligion.html*.

224 Carter, R.M., 2008. Knock, knock: where is the evidence for dangerous human-caused global warming? *Economic Analysis & Policy* (Journal of the Economic Society of Australia - Queensland) 32, 107-202.

225 Cohen, M., 2009 (10 Dec.). Beyond Debate. Times Higher Education Supplement. *http://www.timeshighereducation.co.uk/story.asp?sectioncode =26&storycode=409454&c=2*.

226 Murdock, P., 2009a. How politicians and the media change our minds. Campaign for Liberty, 7 Oct. 2009. *http://www.campaignforliberty.com/article.php?view=134*.

Murdock, P., 2009b. The climate change propaganda machine. Campaign for Liberty, 12 Dec. 2009. *http://www.campaignforliberty.com/article.php?view=432.*

227 Nova, J., 2009. *Skeptics Handbook II! Global Bullies Want Your Money.* *http://joannenova.com.au/2009/11/skeptics-handbook-ii-global-bullies-want-your-money.*

228 Eight independent assessments or tests of the hypothesis of dangerous human-caused global warming, as covered in the following AEF lecture: Carter, R.M., 2007. *Balance and context in the global warming debate.* Australian Environment Foundation Annual Conference, Melbourne, 7 Sept. 2007. *http://www.youtube.com/watch?gl=DE&hl =de&v=FOLkze-9GcI&feature=related.*

1. Jensen, D., *et al.* 2007. Scientifically expert dissenting report to an Australian Parliamentary Standing Committee on Science and Innovation Inquiry into Geosequestration Technology, tabled 13 Aug. 2007.

The dissenting report concluded '*Climate change is a natural phenomenon that has always been with us, and always will be. Whether human activities are disturbing the climate in dangerous ways has yet to be proven.*' *http://www.aph.gov.au/house/committee/scin/geosequestration/report.htm.*

2. Spencer, R.W., Braswell, W.D., Christy, J.R. & Hnilo, J., 2007. Cloud and radiation budget changes associated with tropical intraseasonal oscillations. *Geophysical Research Letters* 134, L15707, *doi:10:1029/2007GL029698.*

A demonstration of cooling caused by changes in cloud cover associated with 30-60-day-long tropical weather cycles, whereby more solar radiation is reflected by increased low level cloud and more terrestrial heat escapes as a result of decreased high level cloud. These effects are not taken into account by current GCM models and on a decadal scale can cause cooling of up to 75 per cent of the predicted carbon dioxide forced warming.

See also: Lindzen, R.S. *et al.*, 2001. Does the Earth have an adaptive infrared iris? *Bulletin of the American Meteorological Society* 82, 417-32.

3. Schwartz, S., 2007. Heat capacity, time constant and sensitivity of Earth's climate system. *Journal of Geophysical Research* 112, D24S05, *doi:10.1029/2007JD008746.*

Schwartz, S., 2008. Reply to comments by G. Foster *et al.*, R. Knutti *et al.*, and N. Scafetta on 'Heat capacity, time constant, and sensitivity of Earth's climate system'. *Journal of Geophysical Research* 113, D15105, *doi:10.1029/2008JD009872.*

A demonstration that the sensitivity of temperature to a doubling of carbon dioxide is of the order of $1^{\circ}C$ rather than the $3\text{-}6^{\circ}C$ cited by the IPCC.

4. Tsonis, A.A., Swanson, K. & Kravtsov, S., 2007. A new dynamical mechanism for major climate shifts. *Geophysical Research Letters* 34, L13705, *doi:10.1029/2007GL030288.*

An analysis of natural climate oscillations such as the Pacific Decadal Oscillation and North Atlantic Oscillation that demonstrates the occurrence of a periodic re-alignment of the climate system as coupling and decoupling occurs between the various internal oscillations, corresponding to a 'synchronized chaos' pattern of global climate change. Synchronized changes occurred in 1913, 1942 and 1978 (the Great Pacific Climate Shift), and changes in twentieth century temperature that reflect these shifts require no external forcing agent to explain them.

5. Camp, C.D. & Tung, K.K., 2007. Surface warming by the solar cycle as revealed by the composite mean difference projection. *Geophysical Research Letters doi: 10.1029/2007GL030207.*

These authors were the first to demonstrate the occurrence of significant global temperature changes of $\sim0.2^{\circ}C$ warming and cooling in accord with the 11-year solar sunspot cycle, and, for the period 1960 and 2005, a close match between global temperature and total solar insolation.

6. Watts, A., 2009. Is the US surface temperature reliable? *http://scienceandpublicpolicy.org/other/is_us_surface_temp_reliable.html.*

Preliminary results of a US-wide survey of official weather stations, published in fuller detail in 2009, revealed that NOAAs National Climate Data Center is incompetent - for '*Their global observing network, the heart and soul of surface weather measurement, is a disaster. Urbanization has placed many sites in unsuitable locations — on hot black asphalt, next to trash burn barrels, beside heat exhaust vents, even attached to hot chimneys and above outdoor grills*'. The US is claimed to have the best weather recording network in the world, so what price global temperature records like the IPCC's, which have been compiled from weather stations that, on average, are even more inadequate than those described by Anthony Watts?

7. Keenan, D.J., 2007. Wei-Chyung Wang fabricated some scientific claims. *http://www.informath.org/apprise/a5620.htm.* See also *http://climateaudit.org/2007/06/18/did-jones-et-al-1990-fabricate-its-quality-control-claims/.*

The weak correlation between global temperature and human carbon dioxide emissions, such as it is, rests heavily on the post-1979 warming displayed by the CRU's surface temperature graph, as used by the IPCC. Satellite and weather balloon measurements show much less, if any, warming trend over the same period. This has led many independent observers to suggest that the CRU graph is contaminated by the urban heat island effect. Doug Keenan's analysis shows that two key papers cited by the IPCC to discount an urban heat island effect are scientifically flawed beyond redemption. The two relevant papers are (i) Jones, P.D., *et al.*, 1990. Assessment of urbanization effects in time series of surface air temperature over land. *Nature*, 347, 169-172; and (ii) Wang, W.-C., *et al.*, 1990. Urban heat islands in China. *Geophysical Research Letters* 17, 2377-80.

8. Nordhaus, W., 2007. The DICE-2007 model of the economics of climate change. *http://nordhaus.econ.yale.edu/*.

Nordhaus' modelling suggests that the anti-carbon dioxide policies recommended by Nicholas Stern and Al Gore would inflict costs on the world economy of approximately US$27 trillion and $34 trillion respectively. Both proposals imply a carbon tax rising to $300/ton of carbon in the medium-term future and up to $800/ton in 2050. A $700/ton carbon tax in the USA would increase the cost of coal-fired electricity by ~150%, at an overall cost of ~$1.2 trillion to the economy. It has never been likely that voters in Western democracies would agree to support such taxes for no identifiable environmental benefit, as is shown by the collapse of the December 2009 climate summit in Copenhagen, and by the failure of President Obama to advance his carbon dioxide cap-and-trade bill through the US legislature.

229 Lindzen, R.S. & Choi, Y.-S., 2009. On the determination of climate feedbacks from ERBE data. *Geophysical Research Letters* 36, L16705, *doi:10.1029/2009GL039628*.

Soon, W., 2007. Implications of the secondary role of carbon dioxide and methane forcing in climate change: past present and future. *Physical Geography* 28, 97-125.

Spencer, R. W. & Braswell, W.D., 2008. Potential biases in feedback diagnosis from observational data: a simple model demonstration. *Journal of Climate* 21, 5624-8.

230 Roderick, M.L. & Farquhar, G.D., 2002 The cause of decreased pan evaporation over the past 50 years. *Science* 298, 1345-6.

231 Calvo, E., Pelejero, C., De Deckker, P. & Logan, G.A., 2007, Antarctic deglacial pattern in a 30 kyr record of sea surface temperature offshore South Australia, *Geophysical Research Letters* 34, L13707, *doi:10.1029/2007GL029937*.

232 Zanchettin, D., Franks, S.W., Traverso P. & Tomasino M., 2008. On ENSO impacts on European wintertime rainfalls and their modulation by the NAO and the Pacific multi-decadal variability described through the PDO index. *International Journal of Climatology* 28, 995-1006.

McLean, J.D., de Freitas, C.R. & Carter, R.M., 2009. Influence of the Southern Oscillation on tropospheric temperature. *Journal of Geophysical Research, 114. Doi:10.1029/2008JD011637.*

McLean, J.D., de Freitas, C.R. & Carter, R.M., 2010. Censorship at AGU: scientists denied the right of reply. SPPI paper, available at *http://scienceandpublicpolicy.org/originals/censorship_at_agu.html.*

233 Tapscott, M., 2009. Peer-reviewed study from Down Under points to nature as global warming source. *Washington Examiner*, 23 July 2009. *http://www.washingtonexaminer.com/opinion/blogs/beltwayconfidential/ Peer-reviewed-study-from-Down-Under-points-to-nature-as-global-warming-source-51489012.html#ixzz0etS67U5w.*

234 Compo, G.P. & Sardeshmukh, P.D., 2009. Oceanic influences on recent continental warming. *Climate Dynamics* 32, 333-42.

Fall, S., Niyogi, D., Gluhovsky, A., Pielke Sr., R.A., Kalnay, E. & Rochon, G., 2009. Impacts of land use and land cover on temperature trends over the continental United States: Assessment using the North American Regional Reanalysis. *International Journal of Climatology, doi: 10.1002/joc.1996.*

McKitrick, R. & McMichaels, P. 2004. A test of corrections for extraneous signals in gridded surface temperature data. *Climate Research* 26, 159-73.

Pielke Sr., R.A. *et al.*, 2007. Unresolved issues with the assessment of multi-decadal global land surface temperature trends. *Journal of Geophysical Research* 112, D24S08, *doi:10:1029/2006JD008229.*

235 Loaiciga, H.A., 2006. Modern-age buildup of CO_2 and its effects on seawater acidity and salinity. *Geophysical Research Letters 33, L10605, doi:10.1029/2006GL026305.*

236 Cotton, W. R., 2009. Weather and climate engineering. In: Heintzenberg, J. & Charlson, R. (eds.), '*Clouds in the Perturbed Climate System: Their Relationship to Energy Balance, Atmospheric Dynamics, and Precipitation*'. MIT Press, pp. 339-68.

Essenhigh, R.E., 2009. Potential dependence of global warming on the residence time (RT) in the atmosphere of anthropogenically sourced carbon dioxide. *Energy & Fuels* 23, 2773-84.

Essex, C., 2007a. There is no global 'temperature'. *Financial Post* (Canada), 24 June 2007. *http://www.canada.com/nationalpost/financial post/story.html?id=cefee93e-4ca0-4490-a08c-5d799926deee.*

Essex, C., 2007b. Fundamental uncertainties in climate modeling. Supplementary Analysis in: McKitrick, R. *et al.*, *Independent Summary for Policymakers of the IPCC Fourth Assessment Report (Working Group 1)*. Fraser Insitute. *http://www.uoguelph.ca/~rmckitri/research/ispm.html.*

Essex, C., McKitrick, R. & Andressen, B., 2007a. Does a global temperature exist? *Journal of Non-Equilibrium Thermodynamics* 32, 1-27.

Essex, C., Llie, S. & Corless, R.M., 2007b. Broken symmetry and long-term forecasting. *Journal of Geophysical Research*, 112, *doi: 10.1029/2007JD008563.*

Green, K., Ball, T. & Schroeder, S., 2004. *The Science Isn't Settled. The Limitations of Global Climate Models.* In Public Policy Sources, No. 80, pp. 1-32. Occasional Paper, Fraser Institute.

Karner, O., 2002. On non-stationarity and anti-persistency in global temperature signatures. *Journal of Geophysical Research* 107, D20. *http://www.aai.ee/~olavi/2001JD002024u.pdf.*

Kucharski, F. *et al.*, 2009. The CLIVARC20C project: Skill of simulating Indian monsoon rainfall on interannual to decadal timescales. Does GHG forcing play a role? *Climate Dynamics* 33(5), Oct. 2009.

Kukla, G. & Gavin, J., 2004. On the credibility of climate predictions. *Global & Planetary Change* 40, 27-48.

Koutsoyiannis, D., Efstratiadis, A., Mamassis, N. & Christofides, A., 2008. On the credibility of climate predictions. *Hydrological Sciences* 53, 671-84.

Paltridge, G., Arking, A. & Pook, M., 2009. Trends in middle- and upper-level tropospheric humidity from NCEP reanalysis data. *Theoretical and Applied Climatology, doi: 10.1007/s00704-009-0117-x.*

Spencer R.W. & Brasell, W.D. 2008. Potential Biases in Feedback Diagnosis from Observational Data: A Simple Model Demonstration. *Journal of Climate* 21, 5624-8.

Stockwell, D.R.B. & Cox, A., 2009. Structural break models of climatic regime-shifts: claims and forecasts. *arXiv:0907.1650v3 [physics.ao-ph],* 12 July 2009, 14 pp.

Swanson, K.L. & Tsonis, A.A., 2009. Has the climate recently shifted? *Geophysical Research Letters* 36, L06711, *doi:10.1029/2008GL037022.*

237 Popular Technology Net, 2009 (13 Dec.). 500 Peer-Reviewed Papers

Supporting Skepticism of 'Man-Made' Global Warming.
http://www.populartechnology.net/2009/10/peer-reviewed-papers-supporting.html.

238 Schneider, S., 1988. Interview in *Discover* magazine. Extract quoted on website – *http://stephenschneider.stanford.edu/Mediarology/Mediarology Frameset.html*.

239 Christine Stewart, 1998. Environment Minister for Canada, Calgary Herald, 14 Dec. 1998.

240 Kellow, A., 2007. *Science and Public Policy: The Virtuous Corruption of Virtual Environmental Science.* Edward Elgar, 218 pp.

241 Mann, M.E., Bradley, R.S. & Hughes, M.K., 1998 Global-scale temperature patterns and climate forcing over the past six centuries. *Nature* 392, 779-87, *doi: 10.1038/33859.*

Mann, M.E., Bradley, R.S. & Hughes, M.K., 1999. Northern hemisphere temperatures during the past millennium: Inferences, uncertainties, and limitations. *Geophysical Research Letters* 26, 759-62.

Mann, M.E. & Jones, P.D., 2003. Global surface temperatures over the past two millennia. *Geophysical Research Letters* 30, *doi: 10.1029/2003GL017814.*

242 Mangini, A., Spotl, C. & Verdes, P., 2005. Reconstruction of temperature in the Central Alps during the past 2000 yr from a $\delta^{18}O$ stalagmite record. *Earth & Planetary Science Letters* 235, 741-51.

Cook, E.R., Palmer, J.G., Cook, B.I., Hogg, A. & D'Arrigo, R.D., 2002. A multi-millenial palaeoclimatic resource from *Lagarostrobus colensoi* tree-rings at Oroko Swamp, New Zealand. *Global & Planetary Change* 33, 209-20.

243 In response to critical attacks on his work, Steve McIntyre established the *Climate Audit* blog site in January 2005. The blog discusses and debates matters relating to the statistical analysis of climate records. In marked contrast to many climate change websites, *Climate Audit* is unremittingly analytical, welcomes constructive criticism and eschews *ad hominem* comments. In 2007 it won the Weblogs 'Best Science Blog on the web' award. See: *http://climateaudit.org/*.

244 McIntyre, S. & McKitrick, R., 2003. Corrections to the Mann *et al.* (1998) proxy data base and northern hemispheric average temperature series. *Energy & Environment* 14 (6), 751-71.

McIntyre, S. & McKitrick, R., 2005. Hockey sticks, principal components, and spurious significance. *Geophysical Research Letters* 32: L03710, *doi:10.1029/2004GL021750.*

245 Wegman, E.J., 2007. Ad hoc Committee Report on the 'Hockey Stick' Global Climate Reconstruction. US House Energy & Commerce Committee, 92 pp. *http://republicans.energycommerce. house.gov/108/home/07142006_Wegman_Report.pdf.*

246 *Nature* editors rejected, without refereeing, the following letter in correction of the MBH 2004 Corrigendum. Apparently scientific honesty was not a matter of concern to them at the time.

Sir, the Corrigendum[A] by Michael Mann and co-authors on global-scale temperature patterns and climate forcing is of more than usual interest, because the paper which is being corrected[B] played a key role in shaping the Intergovernmental Panel on Climate Change's view[C] that late 20th climate warming was unusual in magnitude.

The last sentence of the Corrigendum reads: 'None of these errors affect our previously published results'. *This statement is incorrect. The errors listed by Mann et al. in fact led to several significant effects, as has been rigorously treated by McIntyre and McKitrick[D]. Most importantly, a re-run of the Mann et al. analysis of proxy climate data with the input errors corrected leads to an output in which late 20th century warming is seen to lie within earlier natural bounds[D]. The (corrected) Mann et al. graph shows that the northern hemisphere temperature index attained its highest values in the early 15th century, and that the 20th century warming cycle has so far only equalled a secondary warm peak that occurred late in the 15th century.*

Given the great public concern regarding the possibility of anthropogenic climate change, scientific journals, and their authors and referees, need to address these matters with punctilious honesty.

Bob Carter (Marine Geophysical Laboratory, James Cook University, Townsville, Qld. 4811)

A. Mann, M.E., Bradley, R.S. & Hughes, M.K., 2004. Corrigendum. Global-scale temperature patterns and climate forcing over the past six centuries. *Nature* 430, 105.

B. Mann, M.E., Bradley, R.S. & Hughes, M.K., 1998. Global-scale temperature patterns and climate forcing over the past six centuries. *Nature* 392, 779-87.

C. Intergovernmental Panel on Climate Change (IPCC), 2001. *Climate Change 2001, Third Assessment Report* (Houghton, J.T. *et al.*, eds.), Cambridge University Press, 881 pp.

D. McIntyre, S. & McKitrick, R., 2003. Corrections to the Mann *et al.* (1998) proxy data base and northern hemispheric average temperature series. *Energy & Environment* 14, 751-71.

247 McIntyre, S. & McKitrick, R., 2010. The M&M Project: Replication Analysis of the Mann *et al.* Hockey Stick.

http://www.uoguelph.ca/~rmckitri/research/trc.html.

Montford, A.W., 2009. *The Hockey Stick Illusion - Climategate and the Corruption of Science*. Stacey International, London, 482 pp.

248 Soon, W. & Baliunas, S., 2003. Proxy climate and environmental changes of the past 1,000 years. *Climate Research* 23, 89-110.

Idso, C., 2010. Medieval Warm Period project. *http://www.co2science.org/data/mwp/mwpp.php.*

249 Williams, P.W., King, D.N.T., Zhao, J.-X. & Collerson, K.D., 2004. Speleothem master chronologies: combined Holocene ^{18}O and ^{13}C records from the North Island of New Zealand and their palaeoenvironmental interpretation. *The Holocene* 14, 194-208.

250 von Storch, Hans, *et al.*, 2004. Reconstructing Past Climate from Noisy Data. *Science* 306, 679-82, *doi:10.1126/science.1096109.*

251 Esper, J., Cook, E.R. & Schweingruber, F.H., 2002. Low-Frequency Signals in Long Tree-Ring Chronologies for Reconstructing Past Temperature Variability. *Science* 22, 2250-3, *doi: 10.1126/science.1066208.*

252 Huang, S., Pollack, H.N. & Shen, P.-Y., 2000. Temperature trends over the past five centuries reconstructed from borehole temperatures. *Nature* 403, 756-8.

Beltrami, H., Ferguson, G. & Harris, R.N., 2005. Long-term tracking of climate change by underground temperatures. *Geophysical Research Letters* 32, L19707, *doi:10:1029/2005GL023714.*

Linsley, B.K., Wellington, G.M. & Schrag, D.P., 2000. Decadal sea surface temperature variability in the subtropical South Pacific from 1726 to 1997 A.D. *Science* 290, 1145-8.

253 Quoted in Gelbspan, R., 1997. *The Heat is On.* Perseus Books.

254 Ja, Crystal, 2009 (3 Dec.). Gagged CSIRO scientist resigns. Sydney Morning Herald. *http://news.smh.com.au/breaking-news-national/gagged-csiro-scientist-resigns-20091203-k7ir.html.*

Breen, C., 2009. Censored CSIRO carbon trading paper – compulsory reading for climate movement. *Critical Times, Solidarity Online*, 14 Dec. 2009. *http://www.criticaltimes.com.au/news/censored-csiro-carbon-tradingpaper-compulsory-reading-for-climate-movement/.*

255 Walsh, K. *et al.*, 2002. *Climate Change in Queensland Under Enhanced Greenhouse Conditions*. Final CSIRO Report for the Queensland Government, 1997-2002, 84 pp.

256 Lindzen, R., 2006. Climate of Fear. Global-warming alarmists intimidate dissenting scientists into silence. *Wall Street Journal*, Opinion, 12 April

2006. *http://www.opinionjournal.com/extra/?id=110008220.*

257   Carter, R.M., 2008. Wong's climate paper clouded by mistakes.
      http://www.theage.com.au/business/wongs-climate-paper-clouded-with-mistakes-20080728-3md1.html.

In preparation for emissions trading legislation, Australian climate Minister Penny Wong published an astonishing green paper in response to what she perceived to be the threat of global warming. The first sentence of the opening section of her paper, entitled 'Why we need to act', contains seven scientific errors — almost one error for every two words. Here is the sentence:

Carbon pollution is causing climate change, resulting in higher temperatures, more droughts, rising sea-levels and more extreme weather.

And here are the errors. First, the debate is not about carbon, but, entirely differently, human carbon dioxide emissions and their potential effect on climate. It makes no more sense for Minister Wong to talk about carbon in the atmosphere than it would for David Milliband to talk about hydrogen comprising most of London's water supply. Use of the term carbon in this way is, of course, a deliberate political gambit, derived from the green ecosalvationist vocabulary and intended to convey a subliminal message about 'dirty' coal.

Second, carbon dioxide is not a pollutant but a naturally occurring, beneficial trace gas in the atmosphere. For the last few million years, the Earth has existed in a state of relative carbon dioxide starvation compared with earlier periods in its history. No empirical evidence exists that levels double or even treble today's will be harmful, climatically or otherwise. Indeed, a trebled level is roughly what commercial greenhouse tomato growers aim for in order to enhance the growth of their product. As a vital element in plant photosynthesis, carbon dioxide is the basis of the entire planetary food chain, literally being the staff of life. Its increase in the atmosphere leads primarily to the greening of the planet. To label carbon dioxide a 'pollutant' is an abuse of language, logic and science.

Third, that enhanced human carbon dioxide emissions are causing dangerous global warming (*'carbon pollution is causing climate change'*) is an interesting and important hypothesis. Detailed consideration of its truth started with the formation of the UN's Intergovernmental Panel on Climate Change in 1988, since when Western nations have spent approaching $100 billion on research into the matter. Despite all the protestations of the IPCC, 20 years on the result has been a failure to identify the human climate signal at global (as opposed to local) level. Accordingly, independent scientists have long since concluded that the most appropriate null hypothesis is that the human global signal lies submerged within the noise of natural climate variability. In other words, our interesting initial hypothesis has turned out to be wrong.

Fourth, the specific claim that carbon dioxide emissions are causing temperature increase is intended to convey the impression that the phase of gentle (and entirely unalarming) global warming that occurred during the late 20th century continues today. Nothing could be further from the truth, in that all official measures of global temperature show that it peaked in 1998 and has been declining since at least 2002. And this in the face of an almost 5 per cent increase in atmospheric carbon dioxide since 1998. Spot the problem?

Fifth, sixth and seventh, the statement that human carbon dioxide emissions will cause *'more droughts, rising sea-levels and more extreme weather'* is unbridled nonsense. Such confident predictions are derived from unvalidated, unsuccessful computer models that even their proponents agree cannot predict the future. Rather, a model projection represents just one chosen virtual reality future out of the many millions of alternatives that could have been generated instead. Complex climate models are in effect sophisticated computer games, and their particular outputs are to a large degree predetermined by the predilections of the programmers. It cannot be overemphasized, therefore, that computer climate projections or scenarios are not evidence, and nor are they suitable for use in environmental or political planning. Moving from virtual reality to real observations and evidence, many of the manifestations of living on a dynamic planet that are cited as evidence for global warming are, at best, circumstantial. The currently observed rates of sea-level change, for example, fall well within the known natural range of past sea-level changes. Should we adapt to the rise? Of course. Should we attempt to 'stop climate change' in order to possibly moderate the expected sea-level rise? Of course not, for one might as well try to stop the clouds scudding across the sky; indeed, that is exactly what one would be trying to do.

The first sentence of the *'Why We Need To Act'* section of the Green Paper is followed by five further short paragraphs that are similarly riddled with science misrepresentation and error. In essence, the section reads like a policy manual for green climate activists, and it represents a parody of our true knowledge of climate change. Such enviro-tripe is profoundly out of touch with the balanced facts of real world science, and is provided in endless quantities to politicians the world over through advisory channels that are clogged with rent seekers, special pleaders and green activists.

258 Pielke, Jr., R.A., 2007. Mistreatment of the economic impacts of extreme events in the Stern Review Report on the Economics of Climate Change. *Global Environmental Change* 17, 302-10.

259   Byatt, I., .Castles, I., Goklany, I.M., Henderson, D., Lawson, N., McKitrick, R., Morris, J., Peacock, A., Robinson, C. & Skidelsky, R., 2006. The Stern Review: A Dual Critique. Part II: Economic Aspects. *World Economics* 7, 165-232.

Nordhaus, W., 2007. Critical Assumptions in the Stern Review on Climate Change. *Science* 317, 201-2.

Tol, R., 2007. A Stern Reply to the Reply to the Review of the Stern Review. *World Economics* 8, 153-9.

260   Lawson, N., 2009. Time for a Climate Change Plan B. *Wall Street Journal*, 21 Dec. 2009. *http://online.wsj.com/article/SB100014240527 48704107604574607793378860698.html*.

261   Gray, R., 2010. Stern Report Was Changed After Being Published. *The Sunday Telegraph*, 31 Jan. 2010. *http://www.telegraph.co.uk/earth/ environment/climatechange/7111618/Stern-report-was-changed-afterbeing-published.html*.

262   Aldous, P., 2001. Breaking the mould. News Feature. *Nature* 410, 13, *doi:10.1038/35065254*.

263   Kaiser, J., 2001. 17 National Academies Endorse Kyoto. News of the Week, *Science* 292, 1275-7, *doi:10.1126/science.292.5520.1275b*.

264   ClimateScienceWatch, 2006. UK Science Academy letter tells EXXONMobil to stop funding global warming denial machine. *http://www.climatesciencewatch.org/index.php/csw/details/royal-societyexxon-letter/*.

265   O'Keefe, W. & Kueter, J., 2006 (22 Sept.). Response to the Royal Society's Letter. Marshall Institute. *http://www.marshall.org /pdf/materials/454.pdf*.

266   Scientific opinion on climate change. Wikipedia article at *http://en.wikipedia.org/wiki/Scientific_opinion_on_climate_change*.

267   Lindzen, R., 2008. Climate Science: Is it currently designed to answer questions? Euresis (Associazone per la Promozione e la Diffusione della Cultura e del Lavoro Scientifico) & Templeton Foundation on Creativity and Creative Inspiration in Mathematics, Science, and Engineering. '*Developing a Vision for the Future*', 29-31 Aug. 2008, San Marino Meeting.

268   McCullagh, D., 8 Dec. 2009. Physicists Stick to Warming Claim Post-ClimateGate. *http://www.cbsnews.com/blogs/2009/12/08/taking_liberties /entry5933353.shtml*.

269   CAN International. About CAN (Climate Action Network). *http://www.climatenetwork.org/about-can*.

270 Ereaut, G. & Segnit, N., 2006. Warm Words: How are we telling the climate story and can we tell it better? Institute for Public Policy Research, Covent Garden, London. *http://www.ippr.org.uk/publications andreports/publication.asp?id=485.*

271 Renwick, J., 2007. Reported in: World climate predictors right only half the time. *N.Z. Climate Science Coalition, Scoop Sci-Tech,* 8 June 2007. *http://www.scoop.co.nz/stories/SC0706/S00026.htm.*

272 cf. Booker, C., 2010. British Council gets in on the climate act. UK Telegraph, 13 Feb. 2010. *http://www.telegraph.co.uk/comment/ columnists/christopherbooker/7231466/British-Council-gets-in-on-theclimate-act.html.*

273 Carter, R.M., 2008. '*The 2006 Climate Change and Governance Conference, Wellington, NZ: Hansenism in the cause of "command and control" climate policies*'. New Zealand Centre for Political Research (NZCPR), 7 June 2008, Research Paper. *http://www.nzcpr.com/ Researchpaper_carter(2).pdf.*

274 Soon, W. & Legates, D., 2009. Galileo Silenced Again. *The Heartland Institute, Perspectives,* Nov. 2009. *http://www.heartland.org/full/ 26365/Galileo_Silenced_Again_.html.*

275 Gould League. *http://www.gould.edu.au/.*

276 Australian Broadcasting Corporation. Prof. Schpinkee's Greenhouse Calculator. *http://www.abc.net.au/science/planetslayer/greenhouse_calc.htm.* See also description and discussion at: *http://wattsupwiththat.com/2008/ 05/31/tv-network-tells-kids-when-their-carbon-footprint-says-they-shoulddie/.*

New Zealand. 4 Million Careful Owners. *http://www.mfe.govt.nz/ website/closed-sites/4million.html.* This website has now been closed. Though the replacement pages no longer include such overt children's propaganda as before, they continue to relate an astounding range of IPCC-derived misinfomation about climate change, e.g. *http://www. mfe.govt.nz/issues/climate/index.html.*

NPower. Climate Cops. *http://www.npower.com/climatecops/.*

277 Cribb, J., 2002 (August). *Australasian Science,* p. 38.

278 Doswell, C., 2006 (25 March). Thoughts about TV Tornado 'Documentaries'. *http://www.flame.org/~cdoswell/crock/ crockumentaries.html.*

279 Phillips, M., 2010 (25 April). The global warming scam. *http://www.melaniephillips.com/diary/archives/001153.html.*

280 Michaels, P.J., 2006 (21 Feb.). Hansen's Hot Hype. *The American Spectator,* 21 Feb. 2006. *http://spectator.org/archives/2006/02/21/hansens-hot-hype.*

281 Claus, G. & Bolander, K., 1977, *Ecological Sanity*, David McKay, New York.

282 Crichton, M., 2003. Aliens cause global warming. Speech at
California Institute of Technology, 17 Jan. 2003.
http://www.michaelcrichton.net/speech-alienscauseglobalwarming.html.

283 Advocacy research. One kind of descriptive policy research, carried
out by people who are deeply concerned about certain social problems,
such as poverty or rape. Their studies seek to measure social problems
with a view to heightening public awareness of them and providing a
catalyst to policy proposals and other action to ameliorate the problem
in question. Occasionally, advocacy research studies bend their
research methods in order to inflate the magnitude of the social
problem described, and thereby enhance the case for public action to
address the issue. See Neil Gilbert's article: *'Advocacy Research and
Social Policy'*, *Crime and Justice (1997)*. *http://www.encyclopedia.com
/doc/1O88-advocacyresearch.html.*

284 A recent exception to this occurred during the discussions that
accompanied parliamentary consideration of an emissions trading bill
in Australia. Independent Senator Steve Fielding asked Climate
Minister Penny Wong to provide him with written answers to three
simple questions about the global warming issue. The Minister's reply,
prepared by her Department of Climate Change, drew largely on IPCC
policy advice. After a discussion meeting at which both the Minister
and the Senator were accompanied by their scientific advisers, Senator
Fielding asked his four advisers to provide an independent audit of the
declared policy in support of the answers to his questions. A full
account of the matter, including the due diligence document, can be
found at: *http://joannenova.com.au/global-warming/the-wong-fielding-
meeting-on-global-warming-documents/.*

Senator Fielding's four advisers concluded in summary:

*'As independent scientists, and at the request of Senator Fielding, we
provide preliminary scientific due diligence* [on IPCC advice on global
warming] *in this document.*

*'Our conclusions are (i) that whilst recent increases in greenhouse gases play
a minor radiative role in global climate, no strong evidence exists that human
carbon dioxide emissions are causing, or are likely to cause, dangerous global
warming; (ii) that it is unwise for government environmental policy to be set
based upon monopoly advice, and especially so when that monopoly is
represented by an international political (not scientific) agency; and (iii) that
the results of implementing emissions trading legislation will be so costly,
troublingly regressive, socially divisive and environmentally ineffective that
Parliament should defer consideration of the CPRS bill and institute a fully
independent Royal Commission of enquiry into the evidence for and against*

a dangerous human influence on climate. We add, with respect to (iii), that the scientific community is now so polarised on the controversial issue of dangerous global warming that proper due diligence on the matter can only be achieved where competent scientific witnesses are cross-examined under oath and under strict rules of evidence.'

285 *Laframboise, D., 2010.* More Dodgy Citations in the Nobel-Winning Climate Report. 23 Jan. 2010. *http://nofrakkingconsensus.blogspot.com /2010/01/more-dodgy-citations-in-nobel-winning.html.*

Laframboise, D., 2010. Greenpeace and the Nobel-Winning Climate Report. 28 Jan. 2010. *http://nofrakkingconsensus.blogspot.com/2010/ 01/greenpeace-and-nobel-winning-climate_28.html.*

286 Dietze, P., 2000. IPCC's most essential errors. *http://www.john-daly.com/forcing/moderr.htm.*

Hinderaker, J., Scott Johnson, S. & Mirengoff, P., 2010. It didn't start with Climategate. *http://www.powerlineblog.com/archives/2010 /01/025294.php?format=print.*

Seitz, F., 1996. A major deception on global warming. *Wall Street Journal,* 12 June 1996. Plus editorial 'Cover Up in the Greenhouse', *WSJ,* 11 July 1996. *http://www.sepp.org/Archive/controv/ipcccont/Item05.htm.*

Wasdell, D., 2007. Political corruption of the IPPC Report? [online]. Available from: *http://www.meridian.org.uk/Resources/Global%20 Dynamics/IPCC/index.htm.*

Wojick, D.E., 2001. The UN IPCC's artful bias. Glaring omissions, false confidence and misleading statistics in the Summary for Policymakers. *http://www.john daly.com/guests/un_ipcc.htm.*

287 Bishop Hill Blog, 2010. Hansen's colleague eviscerates AR4 Chapter 9. *http://bishophill.squarespace.com/blog/2010/2/9/hansens-colleague-eviscerates-ar4-chapter-9.html.*

Source of original criticism: *http://pds.lib.harvard.edu/pds/view/7798293?n=17.*

288 Von Storch, H., 2005. Hans von Storch on Barton. *Prometheus,* 8 July 2005. *http://sciencepolicy.colorado.edu/prometheus/archives/climate_change /000486hans_von_storch_on_b.html.*

289 The exception being that the IPCC has announced that it will conduct an investigation into the Climategate affair.

290 Landsea, C., 2005. Chris Landsea leaves IPCC: an open letter to the community. *http://sciencepolicy.colorado.edu/prometheus/archives /science_policy_general/000318chris_landsea_leaves.html.*

291 House of Lords, 2005. The Economics of Climate Change. Select

Committee on Economic Affairs, *2nd Report of Session 2005-06, volume 1: Report (Paper 12-I)*.

292 'Until recently', because politicians in several Western nations, including USA and Australia, discovered in late 2009 that public support had largely evaporated for the carbon dioxide taxation systems, disguised as emissions trading schemes, that they planned to implement; and that evaporation was surely caused in part by the unflagging efforts of independent scientists to inform the public of the true facts of the global warming matter.

293 Sceptical Scientists. The Climate Sceptics Party website. *http://www.climatesceptics.com.au/scepticalscientists/*.

294 McKitrick, R. (ed.), 2007. *Independent Summary for Policymakers*, Fraser Institute. http://*www.fraserinstitute.org/researchandpublications /publications/3184.aspx*.

295 Open Letter to the Secretary-General of the United Nations, and list of Signatories. UN climate conference taking the World in entirely the wrong direction. 13 Dec. 2007. *http://www.nationalpost.com/news/story.html?id=164002; http://www.nationalpost.com/news/story.html?id=164004*

296 Open Letter to the UN Secretary General. Copenhagen Climate Challenge. Scientists ask UN for Convincing Evidence. International Climate Science Coalition, 8 Dec. 2009. *http://www.copenhagenclimatechallenge.org*.

297 Manhattan Declaration on Climate Change. 'Global Warming' is not a global crisis. International Climate Science Coalition, 4 March 2008. *http://www.climatescienceinternational.org/index.php?option =com_content&task=view&id=37&Itemid=54*.

298 US Senate Environment & Public Works Committee Minority Staff Report (Inhofe), 11 Dec. 2008. More than 700 International Scientists Dissent Over Man-Made Global Warming Claims. Scientists Continue to Debunk 'Consensus' in 2008 and 2009. *http://epw.senate.gov/public/index.cfm?FuseAction=Minority.Blogs&Cont entRecord_id=2674e64f-802a-23ad-490b-bd9faf4dcdb7*.

299 Global Warming Petition Project (The Oregon Petition). *http://www.petitionproject.org/*

300 House of Commons Science and Technology Committee, 2006 Scientific Advice, Risk and Evidence Based Policy Making. Seventh Report of Session 2005-06. *http://www.publications.parliament.uk/pa/cm 200506/cmselect/cmsctech/900/900-i.pdf*.

301 Grove, J.M., 1988. *The Little Ice Age*. Routledge, pp. 1-450 (ISBN 0-

415-01449-2).

302 Steffens, M., 2008. Murdoch's warning on 'our national icon': no bludgers wanted. *Sydney Morning Herald*, 3 Nov. *http://www.smh.com.au/news/national/murdochs-warning-on-our-national-icon-no-bludgerswanted/2008/11/02/1225560645017.html*.

303 Marchant, G. & Mossman, K., 2005. Arbitrary and Capricious. The Precautionary Principle in the European Union Courts. International Policy Network, London, 104 pp. *http://www.policynetwork.net/up loaded/pdf/Arbitrary-web.pdf*.

304 Wigley, T.M.L., 1998. The Kyoto Protocol: CO_2, CH_4, and climate Implications. *Geophysical Research Letters* 25, 2285-8.

305 Christy, J., 2009. Hearing on scientific objectives for climate change legislation. The Committee on Ways and Means, US House of Representatives, One Hundred Eleventh Congress (First Session), 25 Feb. 2009. *http://waysandmeans.house.gov/hearings.asp?formmode=view&id=7847*.

306 Beresford, P., 2009. Eco barons lead the way. *Sunday Times*, 1 March. *http://business.timesonline.co.uk/tol/business/specials/article5816774.ece*.

This article provides a startling account of the degree to which rich people have been leading the investment charge into green business areas such as electric cars, solar power and geothermal energy. Beresford estimates that 100 tycoons or wealthy families, each individually worth GBP200 million or more, have invested GBP267 billion into environmental business causes. Six out of the top ten investors together represent GBP80 billion of investment, and are communications or computer titans (Bill Gates, Michael Bloomberg, Michael Otto, Paul Allen, Sergey Brin and Larry Page). That such high profile persons are making such large investments obviously carries a strong political influence.

307 Klaus, V., 2007 (March). Address delivered to the US House Committee on Energy and Commerce. *http://www.thepraguepost.com/articles/2007/03/28/be-afraid.php*.

308 Alley, R.B., *et al.*, 2002. *Abrupt Climate Change. Inevitable Surprises*. US National Academy of Sciences, Comittee on Abrupt Climate Change. National Academy Press, Washington, D.C., 230 pp.

309 Soon, W. & Yaskell, S.H., 2003. Year without a summer. *Astronomical Society of the Pacific, Mercury*, 32.

Stommel, H. & E., 1983. *Volcano Weather: The Story of 1816, the Year without a Summer*, Seven Seas Press, Newport Rhode Island (ISBN 0-915160-71-4).

310 Worster, D., 1979. *Dust Bowl: The Southern Plains in the 1930's*. Oxford

UP, 277 pp.

National Drought Mitigation Center (USA). Drought in the Dust Bowl years. *http://www.drought.unl.edu/whatis/dustbowl.htm*.

311 Chylek, P., Box, J.E. & Lesins, G., 2004. Global warming and the Greenland ice sheet. *Climatic Change* 63, 201-21.

312 Hoffman, D. & Simmons, A., 2008. *Resilient Earth*. Booksurge Llc. (ISBN-10: 143921154X).

313 Abboud, L., 2008. An exhausting war on emissions. *Wall Street Journal*, 30 Sept. 2008. *http://online.wsj.com/article/SB122272533893187737.html*.

314 Lomborg, B., 2008. Global warming: why cut one 3,000th of a degree? *The Times*, 30 Sept. 2008. *http://www.timesonline.co.uk/tol/comment/columnists/guest_contributors/article4849167.ece*.

315 Upton, S., 2008. We need new middle ground on climate change. *The Dominion Post*, 2 Dec. 2008. *http://www.stuff.co.nz/blogs/opinion/743964*.

316 Peel, M.C., Finlayson, B.L. & McMahon, T.A., 2007. Updated world map of the Koppen-Geiger climate classification. *Hydrology & Earth System Sciences* 11, 1633-1644. *http://www.hydrol-earth-systsci.net/11/1633/2007/hess-11-1633-2007.html*.

317 Brunner, R.D. & Lynch, A.H. 2010. Adaptive Governance and Climate Change. *Meteorological Society of America*, 424 pp. (ISBN: 9781878220974).

318 Titus, J. G., *et al.*, 2009. State and local governments plan for development of most land vulnerable to rising sea-level along the US Atlantic coast. *Environmental Research Letters* 4. *doi:10.1088/1748-9326/4/4/044008*. *http://www.iop.org/EJ/article/1748-9326/4/4/044008/erl9_4_044008.pdf?requestid=058698c2-eac8-4b9f-add9-00653751c40a*.

319 Lawson, N., 2009. Time for a Climate Change Plan B. *Wall Street Journal*, 22 Dec. 2009. *http://online.wsj.com/article/SB10001424052748704107604574607793378860698.html*.

320 Sarewitz, D., 2010. World view: Tomorrow never knows. *Nature* (6 Jan. 2010) 463, 24, *doi:10.1038/463024a*. *http://www.nature.com/news/2010/100106/full/463024a.html*.

321 GeoNet, 2006. The GeoNet Project. Monitoring geological hazards in New Zealand [online]. Available from: *http://www.geonet.org.nz/*; *http://www.eqc.govt.nz*.

322 Tracinski, R., 2009. ClimateGate – The Fix is In. Real Clear Politics, 24 Nov. 2009. *http://www.realclearpolitics.com/articles/2009/11/24/the_fix_is_in_99280.html*.

323 'Climategate' – CRU hacked into and its implications.
http://www.bbc.co.uk/blogs/paulhudson/2009/11/climategate-cru-hacked-into-an.shtml.

324 Hudson, P., 2009. What happened to global warming? *BBC One-minute World News*, 9 Oct. 2009. *http://news.bbc.co.uk/2/hi/science/nature/8299079.stm.*

325 East Anglia Confirmed Emails from the Climatic Research Unit - 1255523796.txt (searchable). *http://www.eastangliaemails.com/emails.php?eid=1052&filename=1255523796.txt.*

326 Kinver, M., 2009. Hackers target leading Climatic Research Unit. *BBC One-minute World News*, 20 Nov. 2009. *http://news.bbc.co.uk/2/hi/science/nature/8370282.stm.*

327 Harrabin, R., 2009. Harrabin's Notes: E-mail impact. *BBC One-minute World News*, 24 Nov. 2009. *http://news.bbc.co.uk/2/hi/8377465.stm.*

328 Jeff, Id, 19 Nov. 2009. Leaked FOIA files 62 mb of gold.
http://noconsensus.wordpress.com/2009/11/19/leaked-foia-files-62-mb-of-gold/#more-6188.

McIntyre, S., 19 Nov. 2009. CRU correspondence.
http://climateaudit.org/2009/11/19/cru-correspondence/.

Watts, A., 19 Nov. 2009. Breaking News Story: CRU has apparently been hacked – hundreds of files released. *http://wattsupwiththat.com/2009/11/19/breaking-news-story-hadley-cru-has-apparently-been-hackedhundreds-of-files-released/.*

329 I thank Mr. Rupert Wyndham for providing the details cited, which are contained in a letter dated 12 March 2009, from him to the UK Information Commissioner. The letter requested that the Commissioner direct the BBC to release a list of the attendees at the meeting, which it has refused to do on the grounds that the meeting falls outside the scope of the Freedom of Information Act. The Commissioner denied the request.

See also the subsequent article:

Delingpole, J., 2010. Why the BBC will always be wrong on Climate Change. *The Telegraph*, UK, 20 Jan. 2010.
http://blogs.telegraph.co.uk/news/jamesdelingpole/100023145/why-the-bbc-will-always-be-wrong-onclimate-change/.

330 Index of /climategate/1/FOIA/mail.
http://www.assassinationscience.com/climategate/1/FOIA/mail/.

Opinion Times, 2010. East Anglia Confirmed Emails from the Climatic Research Unit – Searchable. *http://www.eastangliaemails.com/index.php.*

331 Costalla, J.P., 2010. Climategate.
http://www.assassinationscience.com/climategate/.

Ebrahim, M., 2009. ClimateGate: 30 years in the making. *http://jonova. s3.amazonaws.com/climategate/history/climategate_timeline_banner*.

Garneau, Dick, 2010. *CLIMATEGATE 1979-2010*. 'Climategate is like a tsunami'. *http://www.telusplanet.net/public/dgarneau/Climategate.htm*.

Monckton, C., 2009. *Climategate: Caught Green-handed! Cold facts about the hot topic of global temperature change after the Climategate scandal*. SPPI Original Paper, 43 pp. *http://scienceandpublicpolicy.org/originals/climategate.html*.

Nova, Joanne, 2010. The ClimateGate Virus. *http://joannenova.com.au/2009/12/the-climategate-virus/*.

332 Mosher, S. & Fuller, T.W., 2010. *Climategate. The Crutape letters*. Createspace, 186 pp., ISBN 1450512437. *https://www.createspace.com/3423467*.

333 Poneke, 15 Jan. 2010. 13 years of Climategate emails show tawdry manipulation of science by a powerful cabal at the heart of the global warming campaign. *http://poneke.wordpress.com/2010/01/15/gate/*.

334 Sheppard, N., 28 Nov. 2009. Climategate's Michael Mann Being Investigated By Penn State. *http://newsbusters.org/blogs/noel-sheppard/2009/11/28/climategates-michael-mann-be-investigated-pennstate*.

335 Vaughan, A., 2009. Senior civil servant to investigate leaked emails between climate scientists. *The Guardian*, 3 Dec. 2009. *http://www.guardian.co.uk/environment/2009/dec/03/leaked-email-uea-inquiry*.

336 Archer, L., 2009. Liberal leadership spill: Tony Abbott wins. *News.com.au*, 1 Dec. 2009. *http://www.news.com.au/national/liberal-leadership-spill-tony-abbott-wins/story-e6frfkvr-1225805630744*.

337 Associated Press, 2009. United Nations panel to examine evidence in leaked climate email case. *The Guardian*, 4 Dec. 2009. *http://www.guardian.co.uk/environment/2009/dec/04/un-panel-uae-hacked-climateemail*.

338 Webster, B., 2009. Met Office to re-examine 160 years of temperature data. *The Times Online*, 6 Dec. 2009. *http://www.timesonline.co.uk/tol/news/environment/article6945445.ece*.

339 BBC One-minute World News, 6 Jan. 2009. BBC Trust to assess science coverage. *http://news.bbc.co.uk/2/hi/entertainment/8444403.stm*.

340 Watts, A., 2010. John Coleman's hourlong news special 'Global Warming – The Other Side' now online. *http://wattsupwiththat.com/2010/01/14/john-colemans-hourlong-news-special-global-warming-the-*

otherside-now-online-all-five-parts-here/.

Sheppard, M., 2010. Climategate: CRU Was But the Tip of the Iceberg. *American Thinker*, 22 Jan. 2010. *http://www.americanthinker.com /2010/01/climategate_cru_was_but_the_ti.html*.

341 Leake, J. & Hastings, C., 2010. World misled over Himalayan glacier meltdown. *The Sunday Times*, 17 Jan. 2010. *http://www.timesonline.co.uk /tol/news/environment/article6991177.ece*.

342 Sethi, N., 2010. Ramesh turns heat on Pachauri over glacier melt scare. *The Times of India*, 19 January 2010. *http://timesofindia. indiatimes.com/india/Ramesh-turns-heat-on-Pachauri-over-glacier-meltscare/articleshow/5474586.cms*.

343 UK House of Commons Science and Technology Select Committee, 22 Jan. 2009. *http://www.parliament.uk/parliamentary_committees/ science_technology/s_t_pn14_100122.cfm*.

344 Nelson, D., 2010. India forms new climate change body. 4 Feb. 2010, *The Telegraph* (UK). *http://www.telegraph.co.uk/earth/environment/ climatechange/7157590/India-forms-new-climate-changebody.html*.

345 Eschenbach, W., 20 Dec. 2009. Darwin Zero Before and After. *http://wattsupwiththat.com/2009/12/20/darwin-zero-before-and-after/*.

346 Watts, A., 25 Nov. 2009. Uh, oh – raw data in New Zealand tells a different story than the 'official' one. *http://wattsupwiththat.com /2009/11/25/uh-oh-raw-data-in-new-zealand-tells-a-different-story-than-theofficial-one/*.

Watts, A., 27 Nov. 2009. More on the NIWA New Zealand data adjustment story. *http://wattsupwiththat.com/2009/11/27/more-on-the-niwa-new-zealand-data-adjustment-story/*.

347 Warner, G., 2009. US Congress investigates Climategate e-mails: this could be the beginning of the end for AGW. Daily Telegraph, 26 Nov. 2009. *http://blogs.telegraph.co.uk/news/geraldwarner/100017954/ uscongress-investigates-climategate-e-mails-this-could-be-the-beginning-of-the-end-for-agw/*.

348 Financial Times Editorial, 20 Dec. 2009. Dismal outcome at Copenhagen fiasco. *http://www.ft.com/cms/s/0/5b49f97a-ed96-11de-ba12-00144feab49a.html*.

349 Ball, J., 21 Dec. 2009. Summit Leaves Key Questions Unresolved. *http://online.wsj.com/article/SB126133548868199257.html*.

350 *Hindu Times*, 23 Jan. 2009. India, China won't sign Copenhagen Accord. *http://beta.thehindu.com/news/national/article93870.ece?homepage=true*.

351 Peiser, B., 2010. Copenhagen And The Demise Of Green Utopia. *Die*

Weltwoche, 20 Dec. 2009, English translation at: *http://www.thegwpf.org /opinion-pros-a-cons/413-benny-peiser-copenhagen-and-the-demiseof-green-utopia.html*.

352 Kiong, E., 27 July 2006. Scientists want climate commission. *http://www.nzherald.co.nz/technology/news/article.cfm?c_id=5&objectid=1 0393213*.

Carter, R.M. *et al.*, 3 July 2009. Scientists call for Royal Commission into Climate Change. *http://joannenova.com.au/2009/07/scientists-call-for-royal-commission-into-climate-change-science/*.

353 Carter, R.M., De Freitas, C.R., Goklany, I.M., Holland, D. & Lindzen, R.S., 2007. Climate change. Climate science and the Stern Review. *World Economics* 8, 161-82.

354 Links to these and other signed petitions can be found at *http://www.climatesceptics.com.au/scepticalscientists/*.

Index